Crude Oil Refining

This book provides an overview of crude oil refining processes and presents a deep analysis of the current context and challenges imposed on players in the downstream industry. *Crude Oil Refining: A Simplified Approach* covers traditional processes of the refining industry, the impact of current trends, and technological routes available to help these players survive in a highly competitive environment.

FEATURES

- Offers a simplified approach to crude oil refining processes
- Discusses economic information related to the downstream business, including refining margins and profitability
- Introduces newer trends in the industry, such as petrochemical integration, crude-to-chemicals refineries, and renewables coprocessing in crude oil refineries
- Presents the challenges related to these new trends and offers technological solutions to overcome them for profitable and sustainable operations
- Describes how the use of biofuels can minimize the environmental impact of transportation fuel in nations of high demand like Brazil

Offering a contemporary view of current challenges and opportunities in the downstream oil and gas business, this practical book is aimed at readers working in the fields of petroleum and chemical engineering.

Crude Oil Refining
A Simplified Approach

Marcio Wagner da Silva

CRC Press
Taylor & Francis Group
Boca Raton London

CRC Press is an imprint of the
Taylor & Francis Group, an **informa** business

First edition published 2023
by CRC Press
6000 Broken Sound Parkway NW, Suite 300, Boca Raton, FL 33487–2742

and by CRC Press
4 Park Square, Milton Park, Abingdon, Oxon, OX14 4RN

CRC Press is an imprint of Taylor & Francis Group, LLC

© 2023 Marcio Wagner da Silva

Reasonable efforts have been made to publish reliable data and information, but the author and publisher cannot assume responsibility for the validity of all materials or the consequences of their use. The authors and publishers have attempted to trace the copyright holders of all material reproduced in this publication and apologize to copyright holders if permission to publish in this form has not been obtained. If any copyright material has not been acknowledged please write and let us know so we may rectify in any future reprint.

Except as permitted under U.S. Copyright Law, no part of this book may be reprinted, reproduced, transmitted, or utilized in any form by any electronic, mechanical, or other means, now known or hereafter invented, including photocopying, microfilming, and recording, or in any information storage or retrieval system, without written permission from the publishers.

For permission to photocopy or use material electronically from this work, access www.copyright.com or contact the Copyright Clearance Center, Inc. (CCC), 222 Rosewood Drive, Danvers, MA 01923, 978-750-8400. For works that are not available on CCC please contact mpkbookspermissions@tandf.co.uk

Trademark notice: Product or corporate names may be trademarks or registered trademarks and are used only for identification and explanation without intent to infringe.

ISBN: 978-1-032-27212-2 (hbk)
ISBN: 978-1-032-27215-3 (pbk)
ISBN: 978-1-003-29182-4 (ebk)

DOI: 10.1201/9781003291824

Typeset in Times
by Apex CoVantage, LLC

Contents

Preface ix
Acknowledgments xi
Author xiii

Chapter 1 Crude Oil 1

    1.1   Introduction 1
    1.2   Primary Treating Processes of Crude Oil 4
    1.3   Crude Oil Derivatives 7

Chapter 2 Crude Oil Distillation 15

    2.1   Crude Oil Distillation Units 15
    2.2   Vacuum Distillation Section 19

Chapter 3 Thermal Conversion Processes 23

    3.1   Visbreaking Process 23
    3.2   Delayed Coking Process 25
    3.3   Solvent Deasphalting Technologies 30

Chapter 4 Catalytic Conversion Processes 35

    4.1   Fluid Catalytic Cracking (FCC) Units 35
    4.2   Residue Fluid Catalytic Cracking (RFCC) Technologies 39
    4.3   The FCC Catalyst: Converting Residues to Added-Value Derivatives 42
    4.4   Meeting the Market Demand through FCC Optimization 43
    4.5   The Petrochemical FCC Alternative: Raising Competitive Advantage 45
        4.5.1   Propylene Production from FCC 46
    4.6   Catalytic Reforming Technologies 49
        4.6.1   Aromatics Separation Section: Ensuring Maximum Added Value 52
        4.6.2   Improving the Yield of Light Aromatics: Molecular Management 54
    4.7   Naphtha Alkylation Technologies 56
    4.8   Naphtha Isomerization 60
    4.9   Light Olefin Condensation 64
    4.10   Etherification Technologies 66
    4.11   Light Paraffin Dehydrogenation 69

Chapter 5 Hydroprocessing Technologies 73

- 5.1 Naphtha Hydrotreating Technologies 77
  - 5.1.1 Coker Naphtha Hydrotreating 77
  - 5.1.2 FCC Naphtha Hydrotreating Technologies 79
    - 5.1.2.1 A Special Challenge: Diene (Diolefin) Control 81
- 5.2 Diesel Hydrotreating Units 82
- 5.3 Bottom Barrel Hydrotreating Technologies (Residue Upgrading) 84
- 5.4 Atmospheric Residue Desulfurization: A Special Case 88
- 5.5 Hydrocracking Technologies 91
- 5.6 The Hydroprocessing Catalysts 99
- 5.7 Deactivation of Hydroprocessing Catalysts 100

Chapter 6 Lubricating Production Refineries 105

- 6.1 Closing the Sustainability Cycle: Used Lubricating Oil Recycling 110
  - 6.1.1 Used Lubricating Recycling Technologies 111
- 6.2 A Glance over the Brazilian Lubricating Market 114

Chapter 7 Refining Configurations 119

- 7.1 Nelson Complexity Index 125
- 7.2 The Effect of Crude Oil Slate over the Refining Scheme 126
  - 7.2.1 Heavier Crude Oil Processing 126
  - 7.2.2 Light Crude Oil Processing 133

Chapter 8 Hydrogen Production 137

- 8.1 Hydrogen and Syngas Production Routes 137
- 8.2 Renewable Hydrogen Generation Routes: Fundamental Enabler to the Energy Transition 141
- 8.3 Hydrogen Network and Management Actions 142
  - 8.3.1 The Role of Catalytic Reforming Units in the Refineries' Hydrogen Balance 143

Chapter 9 Caustic Treating Processes 147

- 9.1 Caustic Treating Technologies 147
- 9.2 Bender Treating Technologies 152

Contents   vii

Chapter 10 Environmental Processes 155
          10.1   Sour Water Stripping Technologies 155
          10.2   Amine Treating Technologies.. 158
          10.3   Sulfur Recovery Technologies... 160
          10.4   Water and Wastewater Treatment Technologies 165
                  10.4.1   Oily Sewer.. 167
                  10.4.2   Stormwater Sewer 167
                  10.4.3   Domestic Sewage 169
                  10.4.4   Steps of Effluent Treatment...................................... 169
                  10.4.5   API Oil-Water Separator 169
                  10.4.6   Dissolved Air Flotation 169
                  10.4.7   Biological Treatment.. 170
                  10.4.8   Membrane Bioreactors (MBR) 170

Chapter 11 A New Downstream Industry 173
          11.1   What Is Petrochemical Integration? 175
          11.2   More Added Value to the Processed Crude: Integrated
                Refining Schemes 177
          11.3   Crude Oil to Chemicals: Zero Fuel and Maximum Added
                Value 178
                  11.3.1   Available Crude-to-Chemicals Routes 189
                  11.3.2   The Residue Upgrading Technologies in the
                            Integration of Refining and Petrochemical Assets ... 190
                  11.3.3   Closing the Sustainability Cycle: Plastic
                            Recycling Technologies 192
          11.4   Renewables Coprocessing in Crude Oil Refineries 194
                  11.4.1   Biofuel Production in Brazil 195
                  11.4.2   Challenges of Renewables Coprocessing in
                            Crude Oil Refineries... 196
                  11.4.3   The Hydrotreated Vegetable Oil (HVO): An
                            Attractive Route to Reach "Green Diesel"............. 199

Chapter 12 The Propylene Production Gap 203
          12.1   Propylene: A Fundamental Petrochemical Intermediate 204
          12.2   Propylene Production Routes 205
                  12.2.1   The Maximum Olefin Operation Mode................... 207
                  12.2.2   The Petrochemical FCC Alternative 209
                  12.2.3   Steam Cracking Units 211
                  12.2.4   Propane Dehydrogenation.. 211
                  12.2.5   Olefin Metathesis 214
                  12.2.6   Methanol-to-Olefin Technologies (MTO)............... 214

Chapter 13 Gas-to-Liquid Processing Routes 219

        13.1   Gas-to-Liquid Technologies 219
                13.1.1   Available Technologies 221
        13.2   Ammonia Production Process: An Overview 224
                13.2.1   Ammonia Production Technologies:
                              Some Commercial Processes 225

Chapter 14 Business Strategy Models Applied to the Downstream Industry 229

        14.1   Porter's Competitive Forces in the Downstream Industry 230
        14.2   Changing the Focus: More Petrochemicals and Less Fuel 236
                14.2.1   Petrochemical and Refining Integration as a Differentiation Strategy 237

Chapter 15 Corrosion Management in Refining Assets 241

        15.1   Naphthenic Corrosion: General Overview 241
        15.2   Corrosion in Sour Water Stripping Units 243
        15.3   Corrosion Process in Amine Treating Units 246
        15.4   Corrosion Processes in FCC Units 248
                15.4.1   The Petrochemical FCC: Raising Competitive Advantage x Corrosion Attention 250
        15.5   Corrosion Management in Hydroprocessing Units 251

Chapter 16 Energy Management and the Sustainability of the Downstream Industry 257

        16.1   Introduction and Context 257
        16.2   Simple and Available Alternatives to Energy Optimization 258
        16.3   The Impact of the Energy Management on the Greenhouse Gas Emissions 262

Index 265

Preface

Despite new and cleaner energy sources, crude oil is still fundamental to sustaining the economic development of nations and the technological development of society. In recent decades, we have observed an increasing pressure on the crude oil industry to reduce the environmental impact of their processes and derivatives as a fundamental part of global efforts to reach a more efficient, cleaner, and sustainable society.

In this book, we deal with the processes applied to ensure higher added value to crude oil and the refining processes that are currently known as the downstream industry. The current scenario imposes great challenges to the players of the downstream industry, both due to the growing pressure to reduce the environmental footprint of their derivatives and the reduction in the demand for fossil transportation fuels.

Some trends and technologies, such as the electrification of automobile fleet and additive manufacturing, have the potential to destroy demand for crude oil derivatives, and this technological development requires high-quality derivatives, such as the base oils applied to produce lubricants, and these facts put refiners under pressure and squeeze the refining margins, leading players to look for new routes and processes to ensure high added value to processed crude oil. The objective of this book is to review classic crude oil refining processes and present an overview of the current scenario of the downstream industry and how the players can survive in this transitive period of the crude oil refining sector.

Acknowledgments

It's impossible to achieve any good result alone, and this book is no different. In this sense, I would like to start by giving my thanks to God! Without the support I received from my family, especially my wife, Ana Glaucia, and my daughter, Manuela, this book simply would not exist. I'm also grateful to the professionals with whom I had opportunities to exchange experiences and knowledge, especially Mr. Suleyman Ozmen, a real friend and an outstanding professional, as well as Mr. Romain Roux of Axens.

In my developing journey, I had the opportunity to learn at the State University of Maringa (UEM) and the State University of Campinas (UNICAMP), so I would like to express my thanks to all the good professors of the chemical engineering courses of both universities. The construction of this book relied on the contribution of some of the main technology developers to the crude oil refining industry, among them Chevron Lummus Global, Honeywell UOP, Axens, and Haldor Topsoe, as well as some of the most relevant trend/consultancy companies, such as IHS Markit, Wood Mackenzie, International Energy Agency (IEA), and the Catalyst Group companies. For all these companies, I offer my thanks.

Author

Marcio Wagner da Silva, PhD, is a process engineer and stockpiling manager in the crude oil refining industry based in São José dos Campos, Brazil. He earned a bachelor's in chemical engineering at the University of Maringa (UEM), Brazil, and a PhD in chemical engineering at the University of Campinas (UNICAMP), Brazil. Dr. da Silva has extensive experience in research, design, and construction in the oil and gas industry, including developing and coordinating projects for operational improvements and debottlenecking to bottom barrel units. Moreover, he earned an MBA in project management at the Federal University of Rio de Janeiro (UFRJ) in digital transformation at PUC/RS, and he is certified in business from the Getulio Vargas Foundation (FGV).

Recently Dr. da Silva has dedicated his efforts to learning and sharing knowledge about the crude oil refining industry and taking part as an industry adviser to the International Association of Certified Practicing Engineers (IACPE), a member of the advisory board of *The Catalyst Review Magazine* from the Catalyst Group, and a member of the advisory board of the Global Energy Transition Forum, which is strictly committed to minimizing the environmental impact of the energy industry in a realistic and sustainable manner.

1 Crude Oil

1.1 INTRODUCTION

Crude oil is a complex mixture of hydrocarbons, which occur naturally in the earth. Crude oil can be separated into fractions through distillation to achieve the most useful derivatives for society, like fuel and petrochemicals.

Choosing an adequate crude oil slate is among the most relevant decisions of refiners. Refining assets are designed considering a narrow range of characteristics of crude oil to be processed. However, over the useful life of the assets, crude oil slates to be processed can undergo great changes either due to a shortage of crude oil with certain characteristics or by supply difficulties linked to geopolitical issues.

The characterization and classification of different types of crude oil aim to establish its value primarily in relation to reference crudes like Brent and WTI (West Texas Intermediate), as well as define the technological and refining routes to adequate processing. Crude oil consists basically of a mixture of hydrocarbons and associated impurities. These impurities normally refer to sulfur, nitrogen, oxygen, and metals. The concentration of these impurities significantly raises the technological challenges of crude oil processing, leading to a reduction in the crude oil prices according to the concentration of the impurities. The determination of the crude slate to be processed in a refinery is based on a blending of crude oil, aiming to achieve an adequate composition of hydrocarbons and contaminants that allow the processing in a reliable and profitable manner. Figure 1.1 presents the main variables considered in the choice of crude oil slate to be processed in a refining asset.

Some scenarios, such as the discovery of abundant reserves of crude oil with characteristics different from those suitable for a given refining asset, can support the decision of capital investments aiming to adapt the refining assets to the processing of a certain type of crude oil. This fact is common when a refiner is an importer and oil reserves are discovered in the local market.

In relation to hydrocarbons, crude oil contains paraffinic, naphthenic, and aromatic molecules that confer the chemical and physical characteristics of crude oil.

Crude oil can be classified according to the physical and chemical characteristics of the hydrocarbons found in the geological reservoir, one of the most common classifications is the API grade, which is based on the specific gravity of crude oil as described in Equation 1.1.

$$API = \frac{141,5}{\rho} - 131,5 \qquad (1.1)$$

Where ρ = specific gravity of crude oil

Table 1.1 presents an example of crude oil classification based on API. It's important to note that the API grade is a basic classification parameter of crude oil.

FIGURE 1.1 Schematic Representation of the "Blending Space" of Crude Oil

TABLE 1.1
Crude Oil Classification Based on API Grade

Classification	API Grade
Light crude	API > 31,1
Medium crude	22, 3 > API < 31,1
Heavy crude	10,0 > API < 22,3
Extra-heavy crude	API < 10,0

Source: Adapted from Guidelines for Application of the Petroleum Resources Management System, 2011

A very relevant characteristic of oils for refining hardware is naphthenic acidity. Naphthenic acidity is determined based on the amount of KOH required to neutralize 1 gram of crude oil. Normally, a mixture of crude oil is sought in the refinery load so that it does not exceed 0.5 mg KOH/g. Above this reference, the bottom sections of the distillation units can undergo a severe corrosive process, leading to shorter periods of the operational campaign and higher operating costs in addition to problems associated with integrity and safety. Naphthenic acidity is directly linked to the concentration of oxygenated compounds in crude oil that tend to be concentrated in heavier fractions, giving instability and odor to the intermediate currents.

Another relevant characteristic of crude oil is the salt (NaCl) content. The presence of salt in the oil leads to serious corrosion problems, mainly in atmospheric distillation units. The salt content after desalting in atmospheric distillation units is controlled to be below 3 ppm.

Sulfur content is also one of the variables used in the characterization of crude oil in view of their impact on the emissions of harmful gases when using derivatives as fuels. In addition, sulfur compounds increase the polarity of raw oils, leading to stabilization of emulsions and greater difficulties in the desalting process. Normally, oils are classified as high in sulfur when they have levels above 0,5% by weight and low in sulfur below this reference. High-sulfur oils require greater hydrotreating capacity to meet the current environmental requirements for the commercialization of oil products. The presence of contaminants like sulfur, nitrogen, and oxygen is another relevant parameter to classify crude oil and has a great impact on defining the required processes needed to produce the required crude oil derivatives. Normally, the lighter crudes present higher yields of added-value streams like naphtha and diesel and less contaminant content, which lead these crudes to achieve higher prices in the international market. Nowadays, crude oil with low sulfur content tends to be more valuated in the market, especially due to the regulation IMO 2020 (the International Maritime Organization's rule on limiting sulfur emissions).

This regulation established that after 2020, the maximum sulfur content in the maritime transport fuel oil (bunker) is 0,5% (m.m) against the past 3,5% (m.m). The main objective is to reduce the SOx emissions from maritime fleets, significantly decreasing the environmental impact of this business.

Maritime fuel oil, known as bunker, is a relatively low-viscosity fuel oil applied in diesel cycle engines to a ship's movement. Before 2020, the bunker was produced through the blending of residual streams as vacuum residue and deasphalted oil with dilutants like heavy gas oil and light cycle oil (LCO). Due to the new regulation, a major part of the refiners will not be capable of producing low-sulfur bunker through a simple blend.

Due be produced from residual streams with high molecular weight, there is a tendency for contaminants accumulation (sulfur, nitrogen, and metals) in the bunker. This fact makes it difficult to meet the new regulation without additional treatment steps, which should lead to an increased production cost of this derivative and the necessity for modifications in the refining schemes of some refineries.

The first alternative to meet the IMO 2020 is the control of the sulfur content in crude oil that will be processed in the refinery. However, this solution limits the refinery's operational flexibility and restricts crude slate suppliers, which can be a threat in scenarios with geopolitical instabilities and crude oil price volatility.

According to related by Fitzgibbon et al. (2017), just only a small part of crude oil is capable of producing an atmospheric residue that meets the new requirement of the bunker sulfur content.

Due to the limitation in the supply of low-sulfur crudes, the use of residue upgrading technologies aiming to adequate the contaminants contained in the streams applied in the production of the bunker is an effective strategy.

Despite the challenges imposed by IMO 2020, some refiners and crude oil producers are positively exposed to the new regulation, like the Brazilian crudes from pre-salt reserves, Russian Ural reserves, and Britannic North Sea reserves.

The Brazilian pre-salt reserves offer low-sulfur crude oil, with sulfur content varying from 0,3% to 0,67% (in mass). These characteristics of the Brazilian crudes represent a great competitive advantage not only to the downstream sector, but it's important to consider the valuation of these crudes in the market considering the restrictions imposed by IMO 2020. Nowadays, the pre-salt reserves represent the main crude oil source for Brazilian refineries, and the Brazilian downstream sector can produce bunker in compliance with the IMO 2020 since 2019.

Nitrogen content is also a relevant characteristic of crude oil to be considered when choosing castings for processing at refineries. Nitrogen compounds tend to stabilize emulsions, leading to greater difficulties in desalting oil. In addition, they are responsible for imparting chemical instability to derivatives, leading to the formation of polymers and color changes, especially in aviation kerosene. Excess nitrogen compounds can also lead to the deactivation of the acid function of catalysts in deep conversion processes, such as FCC.

The metal content in crude oil is a relevant variable since, like other contaminants, it tends to be concentrated in heavier fractions of oil. These fractions tend to be processed in deep conversion units, such as hydrocracking and catalytic cracking, and they tend to plug the pores of the catalysts, leading to the rapid deactivation of these catalysts, significantly increasing operating costs, and requiring the installation of guard beds to protect the active catalysts.

As previously mentioned, crude oil considered light tends to be more valued in the market, especially in the current market scenario in which there is a tendency to increase the demand for petrochemical intermediaries to the detriment of transportation fuels.

The adequate characterization of crude oil allows for establishing the main challenges for its processing and the mixture of crude oil necessary to reach an adequate cast for each refining hardware, in terms of either profitability or the maximum contaminant levels allowed for reliable processing, and that ensures the integrity of operational assets.

1.2 PRIMARY TREATING PROCESSES OF CRUDE OIL

The reliability of the processing units is fundamental to allow refiners to achieve the desired reliability and keep the competitiveness and the consumer market supply. The operational continuity of a refinery relies on some factors and a strong management system. However, the quality of the raw material (crude oil) is one of the main factors in ensuring the reliability and integrity of the refining processes. Normally, crude oil that will be processed in the refineries must meet some quality requirements aiming to preserve the separation and conversion processing units, mainly the atmospheric distillation unit. The maximum water and sediment content in crude oil is controlled so that it will be lower than 1% in volume. Other relevant parameters are diluted salt content and the total acid number (TAN), which is defined as the quantity of KOH (potassium hydroxide) needed to neutralize 1 gram of crude oil.

To achieve these requirements, crude oil undergoes a series of treatments. This "primary treatment" aims to ensure the life cycle of the downstream and midstream assets. These processes are generally focused on separating water, gas, and oil phases

Crude Oil

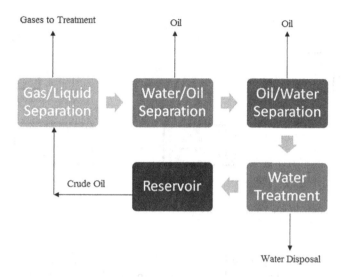

FIGURE 1.2 Steps of the Primary Treatment of Crude Oil

still in the upstream assets. Figure 1.2 shows the basic steps of the primary treatment of crude oil through a block diagram.

Crude oil is drawn from the reservoir, and the separation of gas and liquid phases is carried out through pressure reduction. In the next step, the liquid phase is pumped into a separator drum to promote the separation of oil and water phases by decantation. In this step, only the free water is separated from the oil. A part of the water is emulsified. Subsequently, the mixture undergoes a new treatment step by applying an electrical field and demulsifier addition beyond the heating that aims to reduce the viscosity and allow better phase separation.

The water-oil phase separation is carried out in decantation vessels, which can be two-phase, when it is realized just by the separation of gas and liquid (water + oil) phases, or three-phase, when it involves the separation of free water from oil additionally. Due to the high superficial area, the separation vessels have a normally horizontal configuration. However, in upstream units with great production flow rate oscillations and large sediment content, the vertical configuration is adopted.

In the oil-water-separation step, the emulsion is broken through the application of a high-intensity electrical field that promotes the water droplet polarization and, consequently, decantation. Unlike what occurs in the refineries during the crude desalting process, the electrical treaters used in the upstream assets are low-speed. In this case, the emulsion is fed in the bottom and distributed under a laminar regime to the internals of the separation vessel.

After the separation step, the water is directed to a treatment system. A simplified configuration of a typical water treatment unit is presented in Figure 1.3.

The brine coming from electrostatic treaters is pumped to degassing vessel to remove dissolved gases. After this step, the oily residue is directed to the tank where the phase separation occurs. The aqueous phase is sent to a new treating cycle

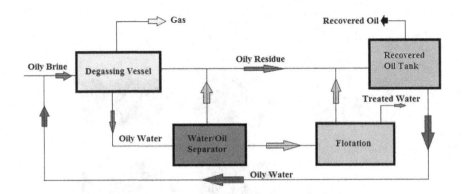

FIGURE 1.3 Oily Water Treatment Process

while the oily phase is pumped to storage. The oily water is directed to a water-oil-separation treatment step, which normally applies API separators. However, in modern sites, hydrocyclones are used due to their higher efficiency. After a flotation step, the treated water can be directed to be disposed of or to be reinjected into the reservoir to improve the recovery of crude oil.

Natural gas produced is directed to treatment steps aiming to reduce the humidity content and sour gases removing. The dehydrating process is carried out through the absorption process with TEG (triethylene glycol), while the sour gases ($H_2S + CO_2$) are removed through amine treatment.

The produced gas stream still undergoes treatment steps aiming to remove heavier compounds (C_3 to C_{5+}) that are considered condensable in natural gas. This process consists basically of the controlled refrigeration of the gas to condense the heavier fractions. The processes generally employed are the Joule-Thomson expansion, simple refrigeration, and turbo expansion. The obtained stream has a great added value and can be applied as a petrochemical feed stream due to its high paraffin content or, according to the consumer market, be directed to the refineries to improve the yield of LPG and gasoline.

As aforementioned, an adequate treatment of crude oil is fundamental to ensure the reliability and availability of the downstream industry. High salt and water content in the crudes leads to higher corrosion and deposition rates in the processing units, reducing the life cycle and increasing operational costs due to unplanned shutdowns. Other assets that experience strong degradation due to the failures in the primary treatment steps are the storage tanks and pipelines. In this sense, the integration between upstream and downstream systems is a key factor in ensuring the sustainability of the crude oil production chain.

When some of the controlled parameters are out of specification, it is necessary to blend different crudes to keep the feed stream to the crude oil distillation unit under controlled conditions. This fact raises the operational costs related to unnecessary operational handling that could be avoided.

Adequate asset management is an important step in the current transformation of the downstream industry. The management system needs to be based on two driving engines: the first is focused on keeping the current operations once they sustain the planned future, and the second is focused on innovative actions to ensure the perenniality of the business. This is an important consideration related to what is called digital transformation. This phenomenon is not only related to technology. Technological advancements make easy access to data possible, but we need a modern and strong management system able to ensure that the right questions will be done to transform these data into information, knowledge, and finally, wisdom.

1.3 CRUDE OIL DERIVATIVES

Crude oil processing produces a series of derivatives with distinct demands and added values. Figure 1.4 presents a simplified process flow diagram for a typical atmospheric crude oil distillation unit and the main derivatives produced in this unit.

The stream considered as fuel gas is normally composed of hydrocarbons in the range C_1 to C_2 and is applied as fuel in the fired heaters and boilers in the own refinery.

The main quality parameters controlled in the fuel gas are the humidity and hydrogen sulfide (H_2S) content. These requirements are normally controlled in dehydration units using propylene glycol and amine treating units, respectively. The concentration of H_2S is controlled to be below 1% in volume and humidity content.

LPG is normally composed of paraffinic and olefinic hydrocarbons in the range of C_3 to C_4 and is applied as domestic and transportation fuel in specific cases. The LPG can contain low quantities of light and heavy hydrocarbons (C_2 and C_5). However, the concentration of these compounds needs to be minimized, aiming not to lose the quality requirements. The concentration of light hydrocarbons is controlled through the Reid vapor pressure (RVP), which is determined by the LPG heating at 37,8°C. The light content is controlled for security reasons, aiming to keep the LPG volatility under safe values to allow storage and handling. The RVP of commercial LPG is controlled to be below 1430 kPa. Once LPG is normally burned into closed environments, the control of burning residue is one of the most important quality requirements of this derivative. The heavy content is controlled through weathering test that evaluates the difficulty in vaporization of LPG. Measuring in an indirect way, the content of C_{5+} in the mixture is normally defined by the boiling temperature of 95% in volume of the mixture under atmospheric pressure and is normally controlled to be below 2°C.

The naphtha streams are normally directed to the refinery gasoline pool according to the refining configuration and the demand of the market where the refiner is inserted. The streams that compose the gasoline pool also depend on the refining scheme. However, it's common for the composition of gasoline pool with straight-run naphtha, cracked naphtha from FCC units, reformed naphtha from catalytic reforming units, isomerized naphtha from isomerization units, and alkylated naphtha produced in catalytic alkylation units.

The gasoline is composed by the blending of these streams containing hydrocarbons with a boiling range of 30–215°C (C_4 to C_{10}). Among the main quality requirements of the gasoline are the antiknock capacity, volatility, corrosivity, pollutants emissions, and the tendency of combustion residue formation in the

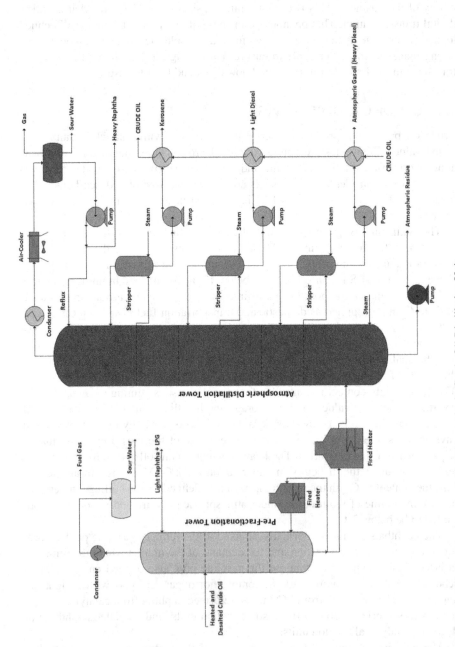

FIGURE 1.4 Typical Process Arrangement of an Atmospheric Crude Oil Distillation Unit

engines. The gasoline antiknock capacity is measured through the octane number that is determined by applying isooctane (2,2,4-trimethyl pentane) as standard with octane number 100 and the n-heptane with octane number 0. The octane number represents a volumetric percentage of isooctane in a mixture with n-heptane, which burns with the same antiknock quality of the analyzed gasoline (measured through sound intensity), the tests to determine the octane number can be the MON (motor octane number) test and the RON (research octane number) test. Common automotive gasoline has an octane number close to 85. The naphtha streams that add higher antiknock capacity to the gasoline are the cracked naphtha from FCC units due to the high olefin content, the reformed naphtha due to the high aromatic concentration, and the naphtha from catalytic alkylation due to the ramified characteristics of the produced kinds of paraffin. However, the aromatic and olefin contents are normally controlled in the final gasoline due to the toxicity and high volatility of these compounds.

The volatility of gasoline is related to the light content in the mixture being directly responsible for the cold starting facility of the internal combustion engines. Gasoline Reid vapor pressure (PVR) indirectly measures the amount of light present in the blend, and for LPG, the gasoline PVR is determined at 37,8°C (100°F) at 1 atm and is usually controlled to be below 55 kPa.

The corrosivity and emissions of the gasoline are controlled through the sulfur content in the final product. Currently, the sulfur content in the gasoline is controlled to be below 50 ppm. For this reason, it's practically impossible to meet this specification without hydrotreating units. Selective hydrotreating units are applied mainly to treat cracked naphtha aiming to reduce the sulfur content with minimum loss of antiknock capacity (due to olefin saturation). The resistance of deposit formation is directly related to the olefin content in the mixture. These compounds are chemically unstable and undergo polymerization, forming polymers that produce deposits and inefficient combustion. The use of antioxidant additives and detergents in the final gasoline can minimize these effects.

A special case of commercialized gasoline is the aviation gasoline that is applied to airplanes equipped with Otto cycle engines. In this case, the hydrocarbons that compose the gasoline have a stricter boiling range (30–170°C) containing ramified paraffin produced by catalytic alkylation processing units.

According to the market to be supplied and the interaction level of petrochemical and refining operations, the light straight-run naphtha can be commercialized as petrochemical naphtha. In this case, it's necessary to guarantee a paraffin content higher than 65%. This alternative tends to be even more applied face with the tendency for a reduction of transportation fuel demand. Furthermore, in markets with high demand by middle distillates, the heavy straight-run naphtha can be directed to compose the diesel or jet fuel pool.

In its turn, jet fuel is a mixture of hydrocarbons between C_5 and C_{15} with a boiling range of 150–300°C; it is applied as fuel to jet turbines, normally applied in aviation. Due to the severity of use conditions, jet fuel has quality requirements quite restricted. The combustion needs to be the cleaner possible to avoid depositions. For this reason, the polyaromatic content is controlled. This is achieved through the smoke point test.

The characteristics of flow under low temperatures are fundamental to jet fuel due to the operational conditions that can achieve temperatures of −50°C. The maximum freezing point for commercial jet fuel is −47°C. For this reason, it's fundamental to ensure an adequate cut point in the distillation step to avoid the drag of heavy paraffin to the intermediate kerosene. The thermal stability is measured through the JFTOT (jet fuel thermal oxidation test), which simulates the operational conditions that the fuel is submitted to.

The corrosivity and chemical stability in relation to the materials applied to the construction of turbines are controlled through the content of total sulfur, mercaptan sulfur, and H_2S. Normally, jet fuel is submitted to a caustic treating step to control these compounds. In modern refining units, this step is carried out in hydrotreating units. The flash point (minimum 40°C) and the electric conductivity are other requirements directly related to the security in the derivative handling.

Diesel is a crude oil derivative that had the most increased demand in the last decades. This derivative is mainly used as a transportation fuel by vehicles equipped with diesel cycle engines and is composed of hydrocarbons between C_{10} and C_{25} with a boiling range of 150–380°C. The diesel ignition quality is measured through the cetane number that corresponds to a volumetric percentage of cetane (n-hexadecane) in a mixture with heptamethylnonane, which burns with the same ignition quality as the analyzed diesel. The linear paraffinic hydrocarbons are the compounds that most contribute to the diesel ignition quality, raising the cetane number while the presence of aromatics reduces this parameter and harms the ignition quality. Currently, the minimum cetane number of commercial diesel is 48. In some countries, like Brazil, the addition of biodiesel in the final product is mandatory, with a minimum concentration of 10% in volume.

The diesel volatility is controlled, aiming to ensure the cold start performance and safety during the handling. The minimum flash point of 38°C and the temperatures of distillation curve correspondent to 50%, 85%, and 95% recovered in volume are controlled in determined limits to ensure the total vaporization in the working conditions. These parameters limit the quantity of naphtha added to the diesel pool.

Another important parameter controlled in the diesel is the plugging point that aims to control the content of linear paraffin that tends to crystallize under low temperatures harm the fuel supply to the engine. The plugging point is determined according to the weather conditions in the region of application. In Brazil, the plugging point is controlled in the range of 0–10°C.

The diesel emissions control is carried out by managing the fuel density aiming to control the content of heavy compounds, especially polyaromatics. Currently, the density of commercial diesel is controlled in the range of 830–865 kg/m³ to ultra-low-sulfur diesel (ULSD). This parameter is controlled to be below 850 kg/m³. In the last decades, there have been great efforts to reduce the environmental damage produced by diesel burn. Nowadays, environmental regulations require the commercialization of low-sulfur diesel with a maximum sulfur content of 10 ppm. However, in some markets, mainly in developing countries, there is still commercialized diesel with higher sulfur content (500 ppm), but this will change soon. This requirement led to the necessity of refiners to expand their hydrotreating capacity.

The viscosity is also a controlled parameter in the diesel, aiming to ensure an adequate nebulization in the combustion chamber. High viscosities can be bad due to the poor dispersion of the fuel, while low viscosities lead to excessive dispersion. Normally, the diesel viscosity is controlled in a range of 2–5 mm^2/s. The diesel lubricity is measured to control the wear due to the friction of the pieces in contact with diesel and is determined by specific tests. The lubricity and the electric conductivity are directly related to the concentration of polar compounds that are reduced after the hydrotreating step. For ULSD, additives are normally used to correct these parameters.

The control of water content and sulfur, nitrogen, and aromatic compounds aims to avoid the proliferation of microorganisms that lead to the filters plugging and add corrosivity to the derivative, as well as raise the stability of oxidation and deposit formation.

Adequate management of crude oil derivative quality requirements is fundamental to achieving the desired goals of performance, safety, and environmental impact. Ensuring the efficiency and reliability of the process responsible for controlling these parameters is a key factor in achieving competitiveness and sustainability in the refining industry.

The fuel oil formulation is carried out by adding diluents to vacuum residue, aiming to achieve a specified viscosity according to the application. Commonly, diluents applied are the gas oil streams from vacuum distillation or streams from deep conversion units like FCC (light cycle oil) or delayed coking (light and heavy gas oils). In some cases, diesel is applied as diluent. The main quality parameters controlled in the fuel oil production are sulfur content, viscosity, the content of sediments and water, vanadium concentration, flash point, and pour point.

The fuel oil is considered a low-sulfur fuel when the maximum concentration of this contaminant is 1% is mass and high-sulfur fuel when the maximum sulfur concentration is 2,5%. The sulfur content control aims to impose a limit on the emissions of harmful gases during the derivative burning. The viscosity control in the fuel oil aims to minimize the transfer costs and ensure adequate flow and vaporization in the burners. The kinematic viscosity of industrial fuel oils (measured at 60°C) is controlled in the range of 600–950 mm^2/s.

The limit of water and sediment content aims to minimize the fouling, deposition, and corrosion in the process equipment and damage to the burners. Furthermore, the water presence reduces the calorific value once part of the released energy is applied to vaporize the water and can provoke flame instability. The maximum vanadium content control aims to minimize the effects of the chemical attack of this metal on the refractory of boilers and fired heaters, as well as metallurgic damages. The maximum vanadium content in the fuel oil is 200 ppm. In its turn, the flash point is applied to control the fugitive emissions and add security during the derivative handling, while the pour point aims to ensure the flow under low temperatures. The pour point specification relies on the weather conditions in the application region.

In some cases, it's necessary to mix different fuel oils to meet the quality requirements. In these cases, it's important to consider the compatibility between the fuel oils. Oils from highly paraffinic crudes show chemical incompatibility with oils produced from crude oil with high asphaltene content, once the presence of paraffin precipitates the asphaltenes due to the resin solubilization that stabilizes the asphaltenes in the solution.

Asphalt is considered a residual fraction of crude oil, normally composed of molecules predominantly aromatic. Asphalt is produced from the vacuum residue that is obtained in the bottom of the vacuum tower, as stated earlier, or from the dilution of the asphaltic residue obtained from the solvent deasphalting process.

The main application of asphalt is the composition of road pavements. Among the asphalt quality requirements are consistency, hardness, ductility, thermoplasticity, viscoelasticity, thermal susceptibility, and durability.

The determination of consistency and hardness of the asphalt aims to define the handling capacity of the derivative. This variable is evaluated by the penetration test, which is performed using a standard needle under specific conditions of loading, temperature, and time. The ductility measures the ability of the asphalt to elongate before rupture. This requirement is directly linked to the strength of the material when applied to the pavement composition.

The thermoplasticity and the viscoelasticity are controlled, aiming at the possibility of hot application of the asphalt and the restoration of the properties of the material after cooling. Thermal susceptibility gives the asphalt the ability to withstand temperature variations without the loss of properties such as consistency and ductility. In turn, the durability test is performed under an aggressive atmosphere of exposure to air and heat, and the other properties are subsequently re-evaluated. The asphalt flash point is controlled to be below 235°C to allow safe handling of the derivative.

The marine fuel oils, called bunkers, are produced from the bottom residue of vacuum distillation. These derivatives are applied as fuels to large ships that operate with diesel cycle engines. Thus, despite also being produced from vacuum residue, the bunker oils have quality requirements different and more severe than the industrial fuel oils.

Due to the bunker's use in diesel engines, it is necessary to control the ignition quality of the bunker. This requirement is evaluated indirectly through the CCAI (calculated carbon aromaticity index), which is evaluated from the density and viscosity parameters that are controlled.

Viscosity is an extremely important variable for the bunker since it is directly related to the ease of nebulization of the derivative in the combustion chamber. High-viscosity oils require a higher heating rate before firing. The bunker viscosity is generally controlled between 2 and 11 mm^2/s (measured at 40°C). Another important feature is the pour point of the bunker. This variable depends on the climatic conditions in the region of application since it is related to the capacity to flow at reduced temperatures, and the bunker pour point is normally controlled between −6°C and 6°C.

The density of the commercial bunker is controlled between 877 and 897 kg/m^3, while the minimum flash point is 60°C to limit fugitive emissions and give safety to the handling of the product. The maximum water and sediment content for the commercialization of the bunker is 0,4% by volume to avoid corrosion and waste deposition in equipment and storage tanks.

BIBLIOGRAPHY

1. Abdel-Aal, H.K., Aggour, M., Fahim, M.A. *Petroleum and Gas Field Processing*. 2nd edition, Marcel Dekker, 2003.
2. Fahim, M.A., Al-Sahhaf, T.A., Elkilani, A.S. *Fundamentals of Petroleum Refining*. 1st edition, Elsevier Press, 2010.
3. Fitzgibbon, T., Martin, A., Kloskowska, A. MARPOL Implications on Refining and Shipping Market. December 2017, www.mckinseyenergyinsights.com/insights/marpol-implications-on-refining-and-shipping-markets/.
4. Gary, J.H., Handwerk, G.E., Kaiser, M.J. *Petroleum Refining: Technology and Economics*. 5th edition, CRC Press, 2007.
5. Odey, F., Lacey, M. IMO 2020—Short-Term Implications for the Oil Market. August 2018, www.schroders.com/bg/uk/asset-manager/insights/markets/imo-2020-what-are-the-short-term-implications-for-the-oil-market/.
6. Robinson, P.R., Hsu, C.S. *Handbook of Petroleum Technology*. 1st edition, Springer, 2017.
7. Speight, J.G. *Heavy and Extra-Heavy Oil Upgrading Technologies*. 1st edition, Elsevier Press, 2013.
8. Robinson, P.R., Hsu, C.S. *Petroleum Science and Technology*. 1st edition, Springer International Publishing, 2019.
9. Guidelines for Application of the Petroleum Resources Management System. November 2011, www.spe.org/industry/docs/PRMS_Guidelines_Nov2011.pdf.

2 Crude Oil Distillation

Crude oil, as found in the reservoirs, has few industrial uses. For crude oil to become useful and economically attractive, it is necessary to separate the fractions in products that have specific industrial uses, like fuels (LPG, gasoline, kerosene, diesel, etc.), lubricants, or petrochemical intermediates. To achieve this objective, crude oil is submitted to a series of physical and chemical processes (called refining complex) aim of adding value to the commodity.

In the refining complex, the first and principal process applied to add value to crude oil is distillation.

2.1 CRUDE OIL DISTILLATION UNITS

The crude oil distillation unit defines the processing capacity of the refinery, and normally, the other processing units are sized based on their yields. Figure 2.1 shows a basic process flow diagram for a typical atmospheric crude distillation unit.

Crude oil is pumped from the storage tanks and preheated by hot products that leave the unit in the heat exchanger's battery, then the crude oil stream receives an injection of water aimed to assist the desalting process. This process is necessary to remove the salts dissolved in the petroleum to avoid severe corrosion problems in the process equipment. The desalting process involves the application of an electrical field to the mixture of crude oil and water aims to raise the water droplets dispersed in the oil phase and accelerate the decanting. As the salt solubility is higher in the aqueous phase, a major part of the salts is removed in the aqueous phase effluent from the desalter, called brine. Normally, the petroleum desalting process is carried out at temperatures between 120°C and 160°C, higher temperatures raise the conductivity of the oil phase and impact the phase separation, and this can lead to dragging oil to the brine and result in process inefficiency.

The desalting process involves the mixture of crude oil with water aiming for the dissolution of the salts, considering the higher solubility of these compounds in the aqueous phase. Figure 2.2 depicts a typical desalting process with two separation stages. The salt content in the desalted crude is normally controlled to be below 3 ppm (as NaCl).

In the desalter exit, the desalted oil is heated again by hot products or pumped around and fed into a flash drum. In this equipment, the lighter fractions are separated and sent directly to the atmospheric tower. The main role of this vessel is to reduce the thermal duty needed in the furnace. Then the stream from the bottom of the flash vessel is heated in the fired heater to temperatures close to 350–400°C (depending on crude oil to be processed) and is fed to the atmospheric tower, where crude oil is fractionated according to the distillation range, like the example presented in Table 2.1.

At the exit of the atmospheric tower, the products are rectified with steam aiming to remove the lighter components.

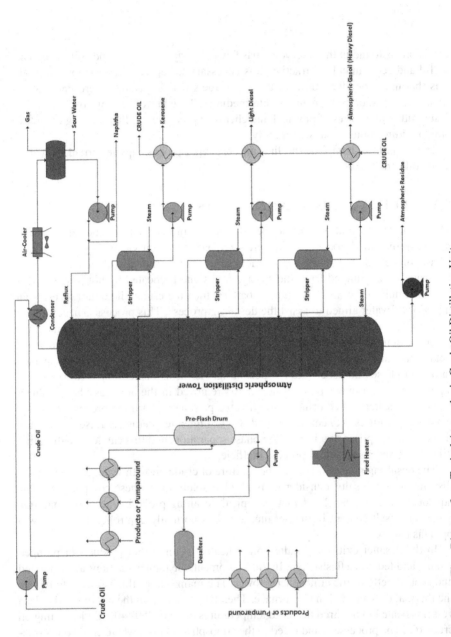

FIGURE 2.1 Process Flow Diagram for a Typical Atmospheric Crude Oil Distillation Unit

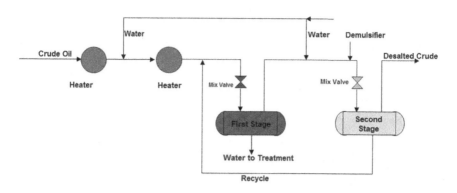

FIGURE 2.2 Crude Oil Desalting Process with Two Separation Stages

TABLE 2.1
Example of Crude Oil Distillation Cuts

Fraction	Distillation Range (°C)
Gases (C_1–C_4)	≤ 30
Light naphtha (C_5–C_7)	30–100
Heavy naphtha (C_8–C_{11})	80–200
Kerosene (C_{11}–C_{12})	170–280
Light diesel (C_{13}–C_{17})	220–320
Heavy diesel (C_{18}–C_{25})	290–350
Atmospheric residue (C_{25+})	350–390

Source: J.H. Gary, G.E. Handwerk, and M.J. Kaiser, *Petroleum Refining: Technology and Economics*, 5th edition, CRC Press, 2007

The gaseous fraction is normally directed to the LPG (C_3-C_4) pool of the refinery and the fuel gas system (C_1-C_2), which will feed the furnaces and boilers. The light naphtha is normally commercialized as a petrochemical intermediate or is directed to the gasoline pool of the refining complex. The heavy naphtha can be sent to the gasoline pool, and in some cases, this stream can be added to the diesel pool since it does not compromise the specification requirements of this product (cetane number, density, and flash point). Kerosene is normally commercialized as jet fuel, while the atmospheric residue is sent to the vacuum distillation tower. In some refining schemes, it's possible to send this stream directly to the residue fluid catalytic process (RFCC) unit. In this case, the contaminant content (mainly metals) of the residue needs to be very low to protect the catalyst of the cracking unit.

Nowadays, faced with the necessity to reduce the environmental impact of the fossil fuels associated with the restrictive legislations, it is difficult for straight-run products to be commercialized directly. The streams are normally directed to the hydrotreating units aimed at reducing contaminant content (sulfur, nitrogen, etc.) before being marketed.

In distillation units with higher processing capacity, normally the flash drum upstream of the atmospheric tower is substituted by a pre-fractionation tower. In

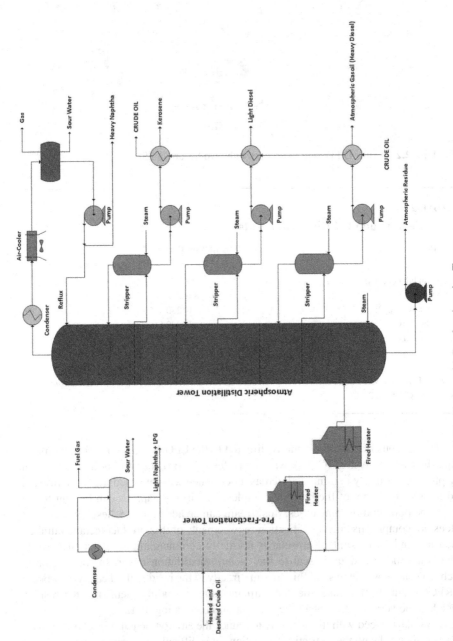

FIGURE 2.3 Typical Arrangement to Atmospheric Distillation with Pre-fractionation Tower

these cases, the main advantage is the possibility of reducing the atmospheric tower dimensions that are implied in cost reductions associated with the unit implementation and improving the hydraulic behavior in the distillation tower, consequently with better fractionation. This arrangement is shown in Figure 2.3.

2.2 VACUUM DISTILLATION SECTION

The bottom stream of the atmospheric column (atmospheric residue) still contains recoverable products capable of being converted into high-added-value derivatives. However, under the process conditions of the atmospheric unit, the additional heating led to thermal cracking and coke deposition.

Aiming to minimize this effect, the atmospheric residue is pumped into the vacuum distillation column, where the pressure reduction leads to a reduction in the boiling point of the heavy fractions allowing the recovery while minimizing the thermal cracking process.

The vacuum generated in the column can be humid, semi-humid, and dry. The humid vacuum occurs when steam injection is applied in the fired heater and in the column aiming to reduce the partial pressure of the hydrocarbons, improving the recovery, while in the semi-humid vacuum, the steam is injected only in the fired heater, minimizing the residence time, reducing the coke deposition. The dry vacuum does not involve the steam injection; in this case, it is possible to achieve pressures between 20 and 80 mmHg, while in the humid vacuum, the column operates under pressures varying between 40 and 80 mmHg. However, it's possible to achieve comparable yields through the injection of stripping steam. Figure 2.4 presents a process arrangement for a typical vacuum generation system in a vacuum crude oil distillation unit.

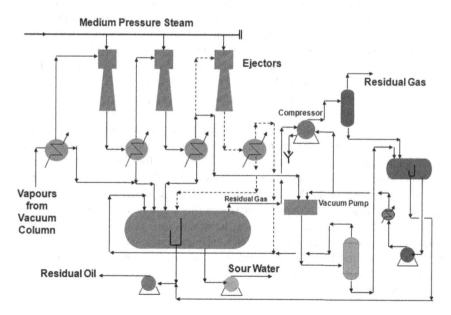

FIGURE 2.4 Process Arrangement for a Typical Vacuum Generation System for a Vacuum Crude Oil Distillation

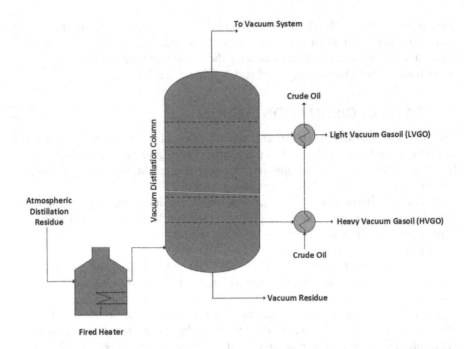

FIGURE 2.5 Schematic Process Flow Diagram for Vacuum Distillation

As shown in Figure 2.5, the traditional arrangement of vacuum units presents two sides drawn: heavy and light gas oil. These streams are normally directed to conversion units like hydrocracking or FCC, according to the adopted refining scheme. The fractionating quality achieved in the crude oil vacuum distillation column has a direct impact on the reliability and conversion unit operation life cycle. Once, in this step, the metal content and the residual carbon (CCR) concentration in the feedstock for these processes are controlled, high values of these parameters lead to quick catalyst deactivation, raising operational costs and reducing profitability.

Some refiners include additional side withdrawals in the vacuum distillation column. When the objective is to maximize the diesel production, it's possible to add a withdrawal of a stream lighter than light vacuum gas oil that can be directly added to the diesel pool or after hydrotreating, according to the sulfur content in the processed crude oil. When crude oil presents high metal content, it's possible to include a withdrawal of fraction heavier than the heavy gas oil called residual gas oil or slop cut. This additional cut concentrates the metals in this stream and reduces the residual carbon in the heavy gas oil, minimizing the deactivation process of the conversion process catalysts as aforementioned. The vacuum residue is normally directed to produce asphalt and fuel oils. However, in most modern refineries, this stream is sent to bottom barrel units as delayed coking and solvent deasphalting to produce higher-value products.

Crude Oil Distillation

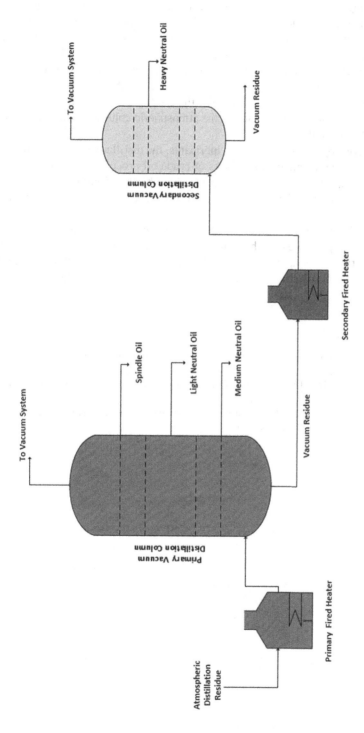

FIGURE 2.6 Vacuum Distillation Process to Produce Lubricants

According to the refining scheme, the installation of vacuum distillation units can be dispensed. Refiners that rely on RFCC units can be sent to the atmospheric residue directly to feed the stream of these units. However, it's necessary to control the contaminant content (metals, sulfur, nitrogen, etc.) and residual carbon (CCR), aiming to protect the catalyst. This fact restricts the crude oil slate that can be processed, reducing the refiner's operational flexibility. On the other hand, in refineries that process extra-heavy crudes, normally the crude oil distillation unit is restricted to the vacuum unit once the yields of the atmospheric column are very low and the coking risk very high.

In refineries optimized to produce lubricants, the distillation process is modified, faced with the paraffinic characteristics of crude oil processed, mainly the vacuum distillation step. The necessity to separate the lubricant fractions requires higher fractionation quality in the column, and some configurations rely on two columns, as presented in Figure 2.6.

The distillation unit design is strongly dependent on the characteristics of crude oil that will be processed by the refinery. For extra-heavy oils, normally the crude is fed directly to the vacuum column. The design is generally defined based on a limited crude oil range that can be processed in the hardware (contaminant content, API grade, etc.).

BIBLIOGRAPHY

1. Speight, J.G. *Heavy and Extra-Heavy Oil Upgrading Technologies.* 1st edition, Elsevier Press, 2013.
2. Robinson, P.R., Hsu, C.S. *Handbook of Petroleum Technology.* 1st edition, Springer, 2017.
3. Gary, J.H., Handwerk, G.E., Kaiser, M.J. *Petroleum Refining: Technology and Economics.* 5th edition, CRC Press, 2007.

3 Thermal Conversion Processes

The refining processes known as thermal processes are conversion routes that do not apply catalysts to produce the desired reactions, normally cracking reactions that convert larger molecules into higher-added-value molecules and derivatives.

Due to their characteristics, the thermal cracking processes tend to produce chemically unstable derivatives like olefins due to the absence of hydrogen and the reaction environment, leading to the necessity of a downstream treatment of the intermediate streams before the application into final derivatives.

3.1 VISBREAKING PROCESS

The search for technologies capable of raising the profitability in the oil derivative production has followed the refining industry since its inception. Over the years, the economic and technological development of nations increased the demand for derivatives drastically, whether for heating, energy production, or transportation fuels.

One of the first technologies applied in the attempt to increase the profitability and the yield of the crude oil refining processes was the visbreaking technology.

The visbreaking process consists of thermal cracking under mild conditions and low residence time aimed to minimize coke production during the feed stream heating step. Figure 3.1 shows a basic process flow diagram for a typical visbreaking processing unit.

The visbreaking process aims to reduce the residual oil viscosity destined to be commercialized as fuel oil, avoiding using high-added-value streams as dilutants (diesel or gas oil) for this purpose.

Typical feedstock to visbreaking units is the vacuum residue from distillation units or, in some cases, reduced crude oil (after the atmospheric distillation step). The feed stream is heated by the fractionating bottom product before being admitted into the fired heater where the thermal cracking reactions are carried out. In the furnace exit, the streams are quenched with gas oil injection aimed at reducing the temperature and stopping the thermal cracking reactions before entering into the distillation column.

In the distillation column, the products are separated, and the light products and LPG are withdrawn from the top. The naphtha is directed to the refinery gasoline pool. A gas oil stream is withdrawn and, normally, is directed to compose the feed stream for FCC unit or hydrocracking process, according to the refining scheme adopted by the refiner. The residual stream from the visbreaking process is directed to fuel oil or asphalt markets, according to the market attended by the refinery.

The process presented in Figure 3.1 is called furnace visbreaking once the thermal cracking reactions occur in the furnace coils.

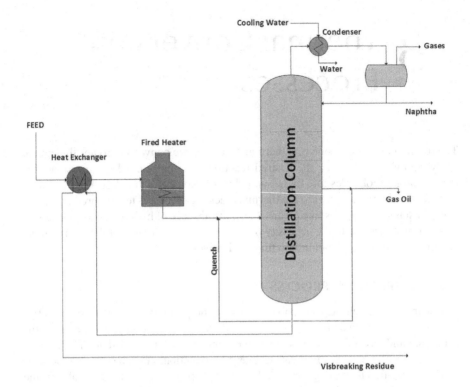

FIGURE 3.1 Typical Arrangement for Visbreaking Processing Units

Another process arrangement for visbreaking processing units is shown in Figure 3.2, called soaker visbreaking.

The main difference between these projects' designs is the presence of a vessel after the furnace, where a major part of thermal cracking reactions occurs. Called a soaker drum, this equipment is applied to enlarge the residence time, which allows operating the furnace with a lower temperature, consequently with a lower probability of coke production. Another advantage is the lower energy consumption in the process.

Another process arrangement to visbreaking units that can be applied is the process where the visbreaking residue is fed to a vacuum distillation column. In this case, the advantage is a higher gas oil yield in the process.

The principal process variables for visbreaking units are the furnace temperature that can vary between 480 and 500°C for the furnace visbreaking and 440 and 460°C for the soaker drum visbreaking. The feed stream pressure can vary from 5 to 35 bar, according to the design concept of the processing unit. The feed stream temperature is normally higher than 300°C. Another relevant process variable is the quench flow rate, applied to interrupt the cracking reactions and achieve an adequate temperature for the fractionating step.

The visbreaking process presents a relatively high operational cost and, once its products are produced from thermal cracking, needs high-cost treatment to be commercialized (normally hydrotreating), which reduces the process's profitability.

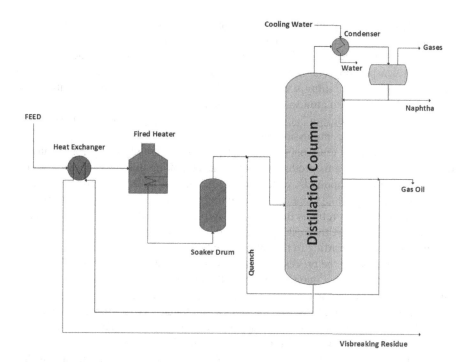

FIGURE 3.2 Visbreaking Process with the Soaker Drum

Furthermore, the process shows a reduced yield (naphtha and gas oil yield between 15% and 25%). The use of visbreaking units was disseminated in the past. However, after the consolidation of vacuum distillation technology, the process fell into disuse by the refiners.

Nowadays, only specific refining schemes apply visbreaking units. An example is the use of this technology together with delayed coking units aiming to produce high-quality coke (needle or anode grade).

Some technology licensors developed research aiming to raise the competitiveness of the visbreaking process as the Shell Soaker Visbreaking technology, licensed by the companies Shell and Lummus. The companies Axens, UOP/Foster Wheeler, and KBR have visbreaking processes in their technology portfolio solutions to the crude oil refining industry.

Despite being an outdated process, as aforementioned, visbreaking technology can be economically attractive in some scenarios and, from the technological point of view, still a very interesting process.

3.2 DELAYED COKING PROCESS

One of the most applied technologies by the refiners is the delayed coking process. Delayed coking employs the thermal cracking concept under controlled conditions to produce light and middle streams (LPG, naphtha, and gas oils) from residual streams, which would normally be used as diluents in fuel oil production.

The typical feed stream for delayed coking units is the residue from the vacuum distillation process that contains the heavier fractions of processed crude oil. However, streams like decanted oil from the FCC unit and asphaltic residue produced in solvent deasphalting can compose the feed stream to the delayed coking unit, depending upon the refining scheme adopted by the refiner. Another possibility is to send the residue from atmospheric distillation directly to the delayed coking unit. In this case, the unit design is quite modified, demanding greater robustness of the fractionating and gas compression section.

Due to the thermal cracking characteristics (low availability of hydrogen during the reactions), the streams produced by the delayed coking unit have a high concentration of olefinic compounds, which are chemically unstable. Furthermore, due to the processing of residual streams that have high contaminant content like nitrogen, sulfur, and metals, the refiners that apply delayed coking units need high hydrotreating capacity to convert these streams into added-value products and that meet the contaminant level according to the environmental regulation.

Figure 3.3 presents the process flow scheme for a typical delayed coking unit.

The feed stream is fed into the bottom of the main fractionating tower, where it is mixed with the heavier fraction of the thermal cracking products and then sent to the fired heater, where thermal cracking reactions are initiated. The reaction conditions are controlled so that the reactions are completed in the coke drums. The residence time in the fired heater must be the lowest possible to minimize the coke precipitation in the fired heater tubes. A manner of minimizing coke formation in the walls of tubes is the steam injection that raises the velocity and consequently reduces the residence time.

After the fired heater, the feed stream is sent to the coke drum or reactor, where the thermal reactions are completed, and the coke is deposited. The thermal cracking

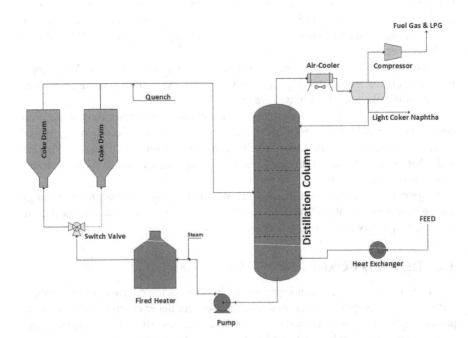

FIGURE 3.3 Typical Arrangement for Delayed Coking Unit

products are removed from the top of the reactor and receive an injection of quench with a cold process stream (normally heavy or middle gas oil) and directed to the main fractionators where the products are separated. The coke deposited in the reactor is removed through a cut with water under high pressure (about 250 bars).

Delayed coking is a process that occurs in batches. In order to make a semi-continuous process, paired reactors are always employed, and for each of the two reactors, one fired heater is used when one reactor is under reaction and the other is in the decoking step. The delayed coking process occurs in cycles that can vary from 14 to 24 hours.

The main operational variables of the delayed coking unit are recycling ratio, which is the quantity of the total feed stream, which corresponds to the heavier fraction of the reaction products that are mixed with the fresh feed; reactor temperature, normally considered in the top of the coke drum; pressure in the top of the reactor; and the time of the reactor cycle.

The recycle ratio normally varies between 5% and 10% (to units dedicated to producing fuels), and the refiners seek to operate the unit with the lower recycle ratio possible in order to maximize the capacity of the plant in processing residual streams. The reactor temperature is close to 430°C and is linked with the fired heater temperature. Throughout the thermal cracking reactions, the temperature falls due to the endothermic characteristics of the reactions.

The pressure in the reactor can vary between 1 and 3,5 bars. In units optimized for producing fuels, the variable is maintained at lower levels. On the other hand, when the unit is dedicated to producing high-quality coke, the unit is operated under higher pressures.

Reactor cycle time is linked to the function performed by the delayed coking in the refining scheme. Units dedicated to producing fuels operate at shorter cycles and units optimized for producing high-quality coke operate under longer cycles.

The coke produced normally is seen as a by-product of the delayed coking unit. However, in some cases, the delayed coking process is optimized to produce high-quality coke, and the coke becomes the principal product of the process.

Depending on the feedstock quality that will be processed, three types of the coke can be produced:

- Shot coke: This is poor-quality coke produced from feedstock with high asphaltene and contaminant (sulfur, nitrogen, and metals) content. Normally, this type of coke is commercialized as fuel.
- Sponge coke: In this case, the feedstock has a lower asphaltene and contaminant content, and the coke can be used as a raw material for the anode production process in the aluminum industry.
- Needle coke: The production of this type of coke requires the processing of feedstock with high aromatic content (decanted oil from FCC, for example), and it is used as a raw material to produce anodes in the steel industry.

As mentioned previously, the production of high-quality coke requires quality control of the feed stream that will be processed. In most cases, the refiners choose to install delayed coking units focusing on the production of middle and light distillates. Therefore, the unit optimization to produce needle coke occurs only in specific cases.

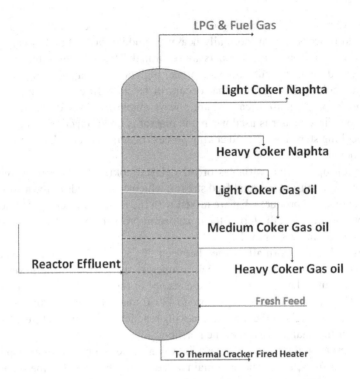

FIGURE 3.4 Main Fractionator Scheme for a Typical Delayed Coking Unit

Figure 3.4 shows a delayed coking main fractionator scheme with the principal process streams.

The heavy gas oil stream is normally directed to the FCC unit or can be utilized as fuel oil. In refining schemes that have deep hydrocracking units, this stream can be used as feedstock to the unit. The sending of this stream to the FCC unit needs to be controlled to avoid the premature deactivation of the catalyst, in the face of a high level of contaminants, mainly nitrogen and metals.

Middle and light gas oils are normally sent to severe hydrotreating units to compose the diesel pool of the refinery. The heavy coker naphtha can be directed like a feed stream to FCC units. When the flash point specification of diesel is not restricted, this stream can be sent to the diesel pool after a deep hydrotreating process.

The lighter fraction of naphtha can be sent to the gasoline pool of the refinery after hydrotreatment or directed to FCC units. In this case, this stream contributes to raising the LPG production in the FCC unit. In some cases, the light coker naphtha can be sent to catalytic reforming units aiming to produce high-octane gasoline or petrochemical precursors (benzene, toluene, and xylene).

The overhead products from the main fractionator are still in the gaseous phase and are sent to the gas separation section. The fuel gas is sent to the refinery fuel gas ring after the treatment to remove H_2S, where it will be burned in fired heaters, while the LPG is directed to treatment and further commercialization.

Figure 3.5 presents a typical scheme for a gas separation section for a delayed conking unit.

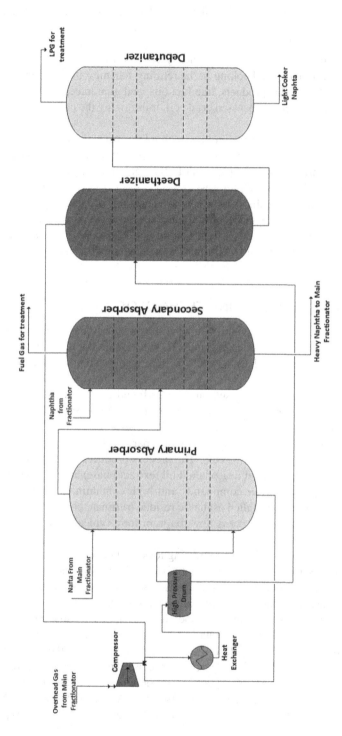

FIGURE 3.5 Basic Process Flow Diagram for a Typical Gas Separation Section from Delayed Coking Unit

The main licensors for delayed coking technology nowadays are the companies Foster Wheeler, ConocoPhillips, Lummus, and KBR.

Delayed coking technology becomes especially attractive for refiners installed in countries with large heavy and extra-heavy crude oil reserves like Mexico and Venezuela. The use of delayed coking in the refining scheme can minimize the production of low-added-value products, like fuel oils, and guarantees higher flexibility to the refinery in relation to processed crude oil, minimizing the necessity to acquire light oils.

On the other hand, the delayed coking technology necessitates the use of high hydrotreatment capacity by refiners once the streams produced by the unity need a severe treating process before being sent to the commercialization. This fact can raise the operational and installation costs.

Another delayed coking disadvantage is the necessity of handling, storage, and commercialization of coke which normally are not the focus of the refiners. Some variations of coking technology as fluid coking and Flexicoking, the last licensed by ExxonMobil, can minimize or eliminate coke formation during the oil residual streams upgrading.

3.3 SOLVENT DEASPHALTING TECHNOLOGIES

The typical feedstock for deasphalting units is the residue from vacuum distillation that contains the heavier fractions of crude oil. The residue stability depends on the equilibrium among resins and asphaltenes once the resins solubilize the asphaltenes, keeping a dispersed phase.

The deasphalting process is based on a liquid-liquid extraction operation where light paraffin (propane, butane, pentane, etc.) is applied to promote resins solubilization, inducing the asphaltene precipitation that corresponds to the heavier fraction of the vacuum residue and concentrates a major part of the contaminants and heteroatoms (nitrogen, sulfur, metals, etc.). The process produces a heavy stream with low contaminant content called deasphalted oil (extract phase) and a stream poor in solvent containing the heavier compounds and high contaminant content, mainly sulfur, nitrogen, and metals, called asphaltic residue (raffinate phase).

Figure 3.6 shows a basic process flow diagram for a typical process deasphalting unit.

The vacuum residue is fed to the extracting tower, where contact with the solvent occurs, leading to saturated compound solubilization. In the sequence, the mixture of solvent/vacuum residue is sent to separation vessels where the separation of asphaltic residue from deasphalted oil occurs, as well as the solvent recovery.

The choice of solvent employed has fundamental importance to the deasphalting process. Solvents that have higher molar mass (higher carbon chain) present higher solvency power and raise the yield of deasphalted oil. However, these solvents are less selective, and the quality of the deasphalted oil is reduced once heavier resins are solubilized, which leads to a higher quantity of residual carbon in the deasphalted oil. Consequently, the contaminant content is raised too. As normally the deasphalting unit aims to minimize the carbon residue, metals, and heteroatoms in the deasphalted

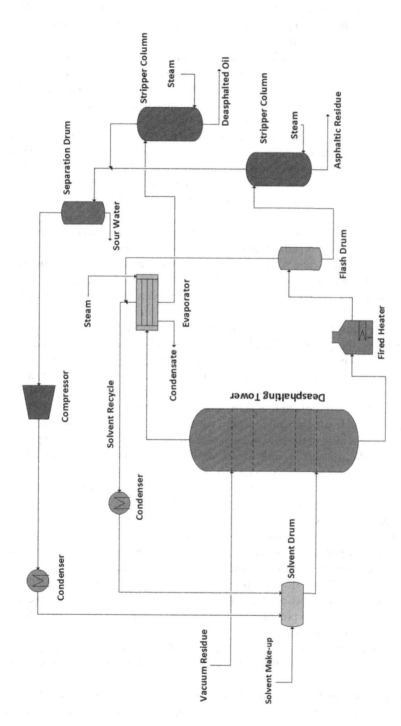

FIGURE 3.6 Typical Arrangement for a Solvent Deasphalting Processing Unit

oil, propane is the usual solvent applied, mainly when the deasphalting process's role in the refining scheme is to prepare feed streams for catalytic conversion processes.

The main operational variables of the deasphalting process are feedstock quality, solvent composition, the relation solvent/feed stream, extraction temperature, and temperature gradient in the extraction tower. Despite being a very important variable, the extraction pressure is defined in the unit design step and is normally defined as the needed pressure to keep the solvent in the liquid phase. In the case of propane, the pressure in the extraction tower is close to 40 bar.

Feedstock quality depends on crude oil characteristics processed by the refinery, as well as the vacuum distillation process. Depending on the fractionating produced in the vacuum distillation unit, the vacuum residue can be heavier or lighter, directly affecting the deasphalting unit yield. Using propane as a solvent, the relation solvent/feed stream is close to 8 and the feed temperature in the extraction tower is close to 70°C.

In refineries focused on fuel production (mainly LPG and gasoline), the deasphalted oil stream is normally sent to the FCC unit. In this case, the contaminant content and carbon residue need to be severely controlled to avoid premature deactivation of the catalyst, which is very sensitive to metals and nitrogen. In refineries dedicated to producing middle distillates, the deasphalted oil can be directed to hydrocracking units.

When the deasphalting process is installed in refining units dedicated to producing lubricants, the quality of deasphalted oil tends to be superior in view that crude oil processed is normally lighter and with lower contaminant content. In this case, the deasphalted oil is directed to an aromatic extraction unit or to hydrotreatment/hydrocracking units. In the last case, the deasphalted oil quality is more critical because of the possibility of premature catalyst deactivation.

The asphaltic residue stream is sent to the fuel oil pool after dilution with lighter compounds (gas oils), or the stream can be used to produce asphalt. Another possibility is to send the asphaltic residue to a delayed coking unit. As the aromatic content in the asphaltic residue is high, the coke produced presents very good quality.

The principal step in the solvent deasphalting process is liquid-liquid extraction, which depends strongly on the solvent properties. In this sense, some licensors developed deasphalting processes based on the solvent in supercritical conditions. Above the critical point, the solvent properties are more favorable to the extraction process, mainly solvency power and the vaporization and compression facility, which reduce power consumption in the process, leading to lower operating costs.

The processes Rose, licensed by KBR; UOP-Demex, licensed by UOP; and Solvahl, licensed by Axens, are examples of deasphalting technologies in supercritical conditions. Figure 3.7 presents a basic process scheme for a typical deasphalting unit under supercritical conditions.

In addition to the quoted processes, the Foster Wheeler, in partnership with UOP, developed the process UOP/FW-SDA, which also applies solvent in supercritical conditions.

As described earlier, the deasphalting process allows added value to residual streams as vacuum residue and, consequently, raises the refiners' profitability.

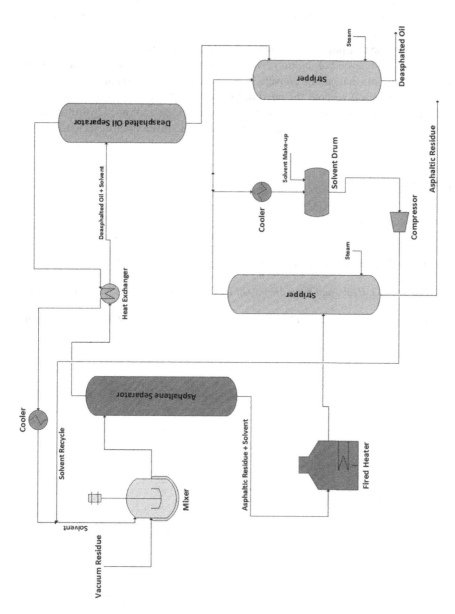

FIGURE 3.7 Typical Arrangement to Solvent Deasphalting Unit under Supercritical Condition

Furthermore, the process can help in the production of higher-quality and cleaner derivates.

As another residue upgrading technology, the deasphalting process raises the refinery flexibility regarding the quality of crude oil processed that can pass to process heavier crude oil that normally has lower cost, and this fact can improve the refining margin.

Currently, the deasphalting technology has lost ground in the more modern refining schemes to delayed coking units since these units can process residual streams, producing streams that can be converted into products with high added value (LPG, gasoline, and diesel) without the need of previous feed stream treatment to remove contaminants. However, the products from delayed coking units need hydrotreatment to be commercialized, which raises the operational and installation costs to the refinery significantly.

BIBLIOGRAPHY

1. Fahim, M.A., Al-Sahhaf, T.A., Elkilani, A.S. *Fundamentals of Petroleum Refining*. 1st edition, Elsevier Press, 2010.
2. Moulijn, J.A. Makkee, M., Van-Diepen, A.E. *Chemical Process Technology*. 2nd edition, John Wiley & Sons Ltd., 2013.
3. Myers, R.A. *Handbook of Petroleum Refining Processes*. 3rd edition, McGraw-Hill, 2004.
4. Speight, J.G. *Heavy and Extra-Heavy Oil Upgrading Technologies*. 1st edition, Elsevier Press, 2013.
5. Robinson, P.R., Hsu, C.S. *Handbook of Petroleum Technology*. 1st edition, Springer, 2017.

4 Catalytic Conversion Processes

As indicated by the name, the catalytic conversion processes apply catalysts to promote the conversion reactions. The catalytic processes can be applied to remove contaminants from final or intermediate streams like the hydrotreating units or to produce new and higher-added-value molecules like hydrocracking, catalytic reforming, and FCC units.

4.1 FLUID CATALYTIC CRACKING (FCC) UNITS

FCC is one of the main processes which give higher operational flexibility and profitability to refiners. The catalytic cracking process has been widely studied over the last decades and has become the principal and most employed process dedicated to converting heavy oil fractions into higher economic value streams.

The installation of catalytic cracking units allows the refiners to process heavier and, consequently, cheaper crude oil, raising the refining margin, mainly in higher crude oil prices scenario, or in a geopolitical crisis, access to light oils can become difficult. The typical catalytic cracking unit feed stream is gas oils from the vacuum distillation process. However, some variations are found in some refineries, like sending heavy coke naphtha, coke gas oils, and deasphalted oils from solvent deasphalting units to processing in the FCC unit.

The catalyst normally employed in FCC units is a solid constituted by small particles of alumina (Al_2O_3) and silica (SiO_2) (zeolite). To the catalyst characteristics and the operational conditions in the catalytic cracking process (temperature higher than 500°C), the process is inefficient for cracking aromatic compounds. Therefore, the more paraffinic the feed stream, higher the unit conversion. Figure 4.1 presents a process scheme for a typical FCC unit.

In a conventional scheme, the catalyst regeneration process consists of the partial carbon burning deposited over the catalyst, according to the following chemical reaction:

$$C + \tfrac{1}{2} O_2 \rightarrow CO$$

The carbon monoxide (CO) is burned in a boiler capable of generating high-pressure steam that supplies others processing units in the refinery.

The principal operational variables in a FCC unit are reaction temperature (normally considered the temperature at the top of the reactor, called a riser), feed stream temperature, feed stream quality (mainly carbon residue), feed stream flow rate, and catalyst quality. Feedstock quality is especially relevant, but this variable is a function of crude oil processed by the refinery, so its difficultly can be changed, but for

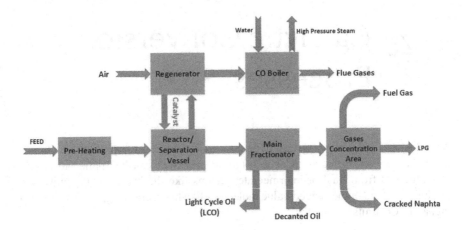

FIGURE 4.1 Schematic Process Flow for a Typical FCC Unit

example, aromatic feedstocks with high metal content are refractory to cracking and conducting to a quick catalyst deactivation.

An important variation of the FCC technology is the RFCC unit. In this case, the feedstock to the process is basically a residue of the atmospheric distillation column. Due to the high carbon residue and contaminants (metals, sulfur, nitrogen, etc.), some adaptations in the unit are necessary, like a catalyst with higher resistance to metals and nitrogen and catalyst coolers. Furthermore, it's necessary to apply materials with the noblest metallurgy due to the higher temperatures reached in the catalyst regeneration step (due to the higher coke quantity deposited on the catalyst). That significantly raises the capital investment in the unit installation. Nitrogen is a strong contaminant to the FCC catalyst because they neutralize the acid sites of the catalyst, which are responsible for the cracking reactions.

When the residue has a high contaminant content, it is common for the feed stream treatment in hydrotreating units to reduce the metal and heteroatom concentration to protect the FCC catalyst.

Typically, the average yield in FCC units is 55% in volume in cracked naphtha and 30% in LPG. Figure 4.2 presents a scheme for the main fractionator of the FCC unit with the principal product streams.

The decanted oil stream contains heavier products and has a high aromatic content. This product is commonly contaminated with catalyst fines, and normally, this stream is directed to be used like fuel oil diluent, but in some refineries, this stream can be used to produce black carbon.

Light cycle oil (LCO) has a distillation range close to diesel, and normally, this stream is directed to treatment in severe hydrotreating units (due to the high aromaticity). After this treatment, the LCO is sent to the refinery diesel pool.

Heavy cracked naphtha is normally directed to the refinery gasoline pool. However, in scenarios where the objective is to raise the production of middle distillates, this stream can be sent to hydrotreating units for further diesel production.

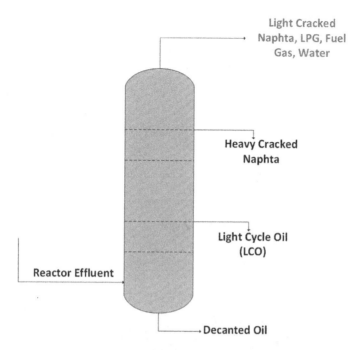

FIGURE 4.2 Main Fractionator Scheme for a Typical FCC Unit

The overhead products from the main fractionator are still in the gaseous phase and are sent to the gas separation section. The fuel gas is sent to the refinery fuel gas ring after treatment to remove H_2S, where it will be burned in fired heaters, while the LPG is directed to treatment (Merox) and further commercialization. The LPG produced by the FCC unit has a high content of light olefins (mainly propylene), so in some refineries, the LPG stream is processed in a propylene separation unit to recover the propylene that has higher added value than LPG.

Cracked naphtha is usually sent to the refinery gasoline pool, which is formed by naphtha produced by other processing units like straight-run naphtha, naphtha from the catalytic reforming unit, and so on. Due to the production process (deep conversion of residues), the cracked naphtha has high sulfur content, and to attend the current environmental legislation, this stream needs to be processed to reduce the contaminant content, mainly sulfur.

The cracked naphtha sulfur removal represents a great technology challenge because it is necessary to remove the sulfur components without molecule saturation that gives a high octane number for gasoline (mainly olefins).

Over the last decades, some technology licensors have developed new processes aiming to reduce the sulfur content in the cracked naphtha with minimum octane number loss. Some of the main technologies dedicated for this purpose are the technology Prime-G+ by Axens, the processes Octagain and SCANfining by ExxonMobil, the process S-Zorb by ConocoPhillips, and the technology ISAL by UOP.

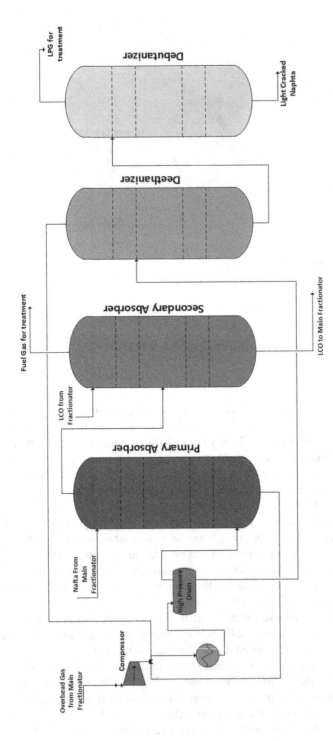

FIGURE 4.3 Basic Process Flow Diagram for a Typical Gas Separation Section from FCC Unit

Usually, catalytic cracking units are optimized to aim for the production of fuels (mainly gasoline). However, some processing units are optimized to maximize light-olefin production (propylene and ethylene). Processing units dedicated for this purpose have their project and operational conditions significantly changed once the process severity is strongly raised in this case.

The reaction temperature reaches 600°C, and higher catalyst circulation rate raises the gas production, which requires a scaling up of the gas separation section. Figure 4.3 presents a typical scheme for a gas separation section for an FCC unit.

In several cases, due to the higher heat necessary for the unit, it is advantageous to operate the regenerator with the total combustion of the coke deposited on the catalyst. This arrangement significantly changes the thermal balance of the refinery once it's no longer possible to resort to the steam produced by the CO boiler.

Over the last decades, the FCC technology has been intensively studied, aiming mainly for the development of units capable of producing light olefins (deep catalytic cracking) and processing heavier feedstocks. The main licensors for FCC technology nowadays are the companies KBR, UOP, Stone & Webster, Axens, Lummus, and Foster Wheeler.

Despite the great operational flexibility that FCC technology gives to the refineries, some new projects have dismissed these units in the refining scheme, mainly when the new refinery objective is to maximize middle distillates products (diesel and kerosene) once this is not the focus of the FCC unit.

4.2 RESIDUE FLUID CATALYTIC CRACKING (RFCC) TECHNOLOGIES

One variation of the FCC that has been widely applied in the last few years is the RFCC. In this case, the feed stream to the process is basically the bottom stream from the atmospheric distillation column, called atmospheric residue, that has high carbon residue and higher contaminant content, like metals, nitrogen, and sulfur.

Due to the feed stream characteristics, the residue catalytic cracking units require design and optimization changes. The higher levels of residual carbon in the feed stream lead to higher temperatures in the catalyst regeneration step and a lower catalyst circulation rate to keep the reactor at constant temperature. This fact reduces the catalyst/oil ratio, which leads to lower conversion and selectivity. To avoid these effects, the RFCC units normally rely on catalyst coolers, as presented in Figure 4.4.

The installation of a catalyst cooler system raises the processing unit profitability through the total conversion enhancement and selectivity for noble products such as propylene and naphtha against gases and coke production. Furthermore, it helps the refinery's thermal balance once high-pressure steam is produced. The use of a catalyst cooler is also necessary when the unit is designed to operate under total combustion mode. In this case, the heat release rate is higher due to the total burn of carbon to CO_2, as presented here:

$$C + \tfrac{1}{2} O_2 \rightarrow CO \text{ (Partial Combustion) } \Delta H = -27 \text{ kcal/mol}$$
$$C + O_2 \rightarrow CO_2 \text{ (Total Combustion) } \Delta H = -94 \text{ kcal/mol}$$

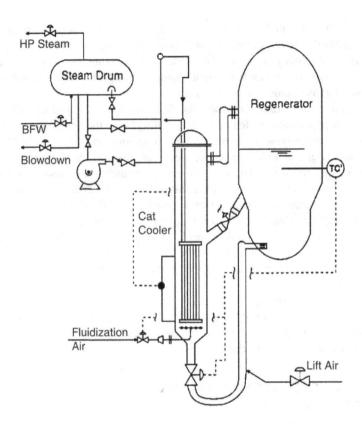

FIGURE 4.4 Catalyst Cooler Process Arrangement for a Typical RFCC Unit

Source: **Handbook of Petroleum Refining Processes, 2004, with Permission**

In this case, the temperature of the regeneration vessel can reach values close to 760°C, leading to higher risks of catalyst damage, which is minimized by installing a catalyst installation. The decision by the total combustion mode needs to consider the refinery thermal balance, once, in this case, will not be possible to produce steam in the CO boiler. Furthermore, the higher temperature in the regenerator requires materials with the noblest metallurgy. This significantly raises the installation costs of these units.

As pointed out earlier, the feed stream characteristics of RFCC units require modifications when compared with the conventional FCC. The presence of higher content of nitrogen compounds leads to an accelerated process of catalyst deactivation through acid site neutralization. The presence of metals like nickel, sodium, and vanadium raises the coke deposition on the catalyst and leads to a higher production of hydrogen and gases. Besides that, it reduces the catalyst life cycle through zeolitic matrix degradation. Beyond these factors, heavier feed streams normally have high aromatic content that is refractory to the cracking reactions, leading to a higher coke deposition rate and lower conversion.

Due to these operation conditions, the RFCC units present higher catalyst consumption when compared with the conventional process. This fact considerably raises the operational costs of the RFCC units. However, the most modern units have applied specific catalysts to process residual feed streams. In this case, the catalyst has a higher porosity aiming to allow a better adaptation to the high aromatic content. Furthermore, the catalyst needs to have a higher metals tolerance.

The control of contaminant content in the feed stream or its effects is a fundamental step of the RFCC process. Sodium content can be minimized through an adequate crude oil desalting process, and the effects of nickel (dehydrogenation reactions) can be reduced by a dosage of antimony compounds, which act like neutralizing agents of the nickel dehydrogenation activity, reducing the generation of low-added-value gases. In turn, vanadium's effects can be controlled through the addition of rare earth to the catalyst, like cerium compounds. The addition of these compounds needs to be deeply studied once it significantly raises the catalyst cost.

The use of visbreaking units to treat the feed streams to RFCC units is a process scheme adopted by some refiners. In these cases, the most significant effect is the reduction in the residual carbon. However, due to its higher effectiveness, the tendency in the last decades has been to treat the bottom barrel streams in deep hydrotreating or hydrocracking units before pumping for RFCC units. With this processing scheme, it's possible to achieve lower contaminant content, mainly metals, leading to a higher catalyst life cycle. Furthermore, hydroprocessing has the advantage of the reduction of the sulfur content in the unit intermediate streams, minimizing the necessity or severity of posterior treatments. A clear disadvantage of this refining scheme is the high hydrogen consumption that significantly raises the operational costs.

Like the conventional FCC units, the main operational variables of RFCC units are the reaction temperature (normally considered the highest point in the reactor, also called the riser), feed stream temperature, feed stream quality, feed stream flow rate, and catalyst quality. It's relevant to say that the conventional FCC units can process atmospheric residue as the feed stream. However, it's necessary to control the contaminant content, mainly metals, which requires processing lighter crudes with higher costs, which raises the operational costs and reduces the flexibility of the refiner in relation to the crude oil supplier.

FCC units have a key role in the current scenario of the downstream industry to allow a closer integration between refining and petrochemical processes in view of the tendency of reduction in the transportation fuel demand, making the petrochemical sector responsible for sustaining the crude oil demand in the next decades.

Beyond the tendency of reduction in transportation fuel demand, the necessity to meet environmental regulations like IMO 2020 requires a strong reduction of contaminant content in residual streams to produce commercial bunker. At first, there is a tendency for the bunker market to be partially supplied by diesel or bottom barrel streams with low sulfur content, leading to a raising in the diesel prices and a devaluation of the high-sulfur-content fuel. This scenario can pressure the refiners with low bottom barrel conversion capacity to carry out capital investments to improve the production of high-added-value derivatives. In this sense, the RFCC technologies can be attractive alternatives to allow a better balance between the flexibility and the quality of processed crude oil (heavier and cheaper crude oils), high yields of petrochemical intermediates, and the production of low-contaminant derivatives, which contribute to enhancing the refining margin.

4.3 THE FCC CATALYST: CONVERTING RESIDUES TO ADDED-VALUE DERIVATIVES

A key factor in the FCC operation is the catalyst applied in the process. The catalyst normally employed in FCC units is a solid constitution of small particles of alumina (Al_2O_3) and silica (SiO_2) (zeolite). To the catalyst characteristics and the operational conditions in the catalytic cracking process (temperature higher than 500°C), the process is inefficient for cracking aromatic compounds. Therefore, the more paraffinic the feed stream, the higher the unit conversion.

The active phase in the FCC catalyst is composed of zeolite, which is responsible for the catalytic activity and selectivity of the catalyst, and of alumina, which is responsible for the cracking of heavier molecules, allowing these molecules to reach the access to the zeolitic phase. The other components of the FCC catalysts are the inert (kaolin) and synthetic matrices that are responsible for mechanical resistance and hardness and act as binder agents between the active phases and the matrix.

According to the process conditions, some compounds can be added to the catalyst with a specific purpose. In refineries that process feed streams with a high amount of nickel, it's common to add antimony that act as a passivator agent. Another deleterious metal is vanadium; in this case, some trap agent is applied to minimize its effects.

As aforementioned, the processing of heavier crude oil leads to more challenging feedstocks to FCC units due to the higher concentration of residual carbon and mainly contaminants such as nickel and vanadium. The nickel acts as a dehydrogenation agent leading to the coke deposition over the catalyst and raises the hydrogen production. Normally, the refiners processing heavier feeds use metal passivators such as boron to keep the deleterious effect of metals under control. The most common method of controlling nickel's effects is to inject antimony into the FCC feed.

Vanadium's effect on the FCC catalyst involves the degradation of the zeolite matrix, leading to the reduction in the catalytic activity, and its action is kept under control through vanadium traps. In the last few years, some catalyst developers are focusing their research on studying the effects of iron in the FCC catalyst. The high concentration of iron is a characteristic of the shale oils produced in North America, and the availability of these crudes increased significantly in the last few years, especially after 2015, when the United States started to export its internal production. Iron is not catalytically active, but this compound can accumulate over the catalyst surface, reducing the porosity, reducing the activity, and leading to dehydrogenation reactions, and it is also a CO promoter. Furthermore, the high concentration of iron can raise the Sox emissions in the catalyst regenerator.

Another dangerous contaminant of the FCC catalyst is sodium. This element promotes an irreversible deactivation of the catalyst through the chemical degradation of the zeolitic matrix. For this reason, adequate control of the crude oil desalting process is fundamental to controlling the sodium content in the FCC feeds, preserving the catalyst life cycle. Nowadays, some refiners inject caustic soda into the crude to improve the desalting characteristics, and stricter control is required in these cases. A less common contaminant found in some crude oil is copper. Its effect is the promotion of dehydrogenation reactions, raising the yield of hydrogen and coke. Copper is present in some NOx-reducing agents.

Aiming to improve the catalytic activity, some developers apply rare earth compounds to the FCC catalysts, such as lanthanum and cerium. These substances significantly raise the activity and selectivity of the final catalyst, but the high cost made the refiners avoid its application. Furthermore, the presence of rare earth in the catalyst improves the yield of gasoline and reduces the light-olefin production in FCC units. In the current scenario, this is exactly the inverse that the refiners are looking for.

The trend of reduction in transportation fuel demand is making refiners optimize their FCC units to maximize petrochemical intermediates against transportation fuels. To achieve this goal, normally the refiners use the most severe conditions such as higher catalyst/oil ratios, higher reaction temperature (TRX), and the use of ZSM-5 as an additive to the catalyst.

The presence of ZSM-5 in the catalyst is capable of improving the yield of light olefins in the FCC unit by up to 8%. One of the most important roles of the refinery optimization teams is to analyze the FCC equilibrium catalysts to find the improvement alternatives based on the contaminant content and the reached conversion of the unit and the degradation observed on the equilibrium catalysts. The volumetric conversion of an FCC unit is defined in Equation (1).

Volumetric Conversion (%) = [Feed − (LCO + Decanted Oil)] / Feed × 100

The fraction LCO and decanted oil (DO) is considered non converted fractions.

The main FCC catalyst developers present in the market nowadays are BASF Catalysts, Albermarle, and W. R. Grace Company.

4.4 MEETING THE MARKET DEMAND THROUGH FCC OPTIMIZATION

According to the market demand, FCC units can be optimized to produce the most demanded derivatives. In traditional FCC units there are normally four operating campaigns:

1. **Maximum Gasoline:** In maximum gasoline campaigns, the processing unit operates under medium or high severity. The severity is limited by the octane number achieved in the cracked naphtha. In refining configurations where the refiners rely on octane boosting units like alkylation, catalytic reforming, and isomerization, there is more flexibility to maximize the gasoline yield in the FCC operating in maximum gasoline mode.

The catalyst formulation to maximum gasoline campaigns involves high zeolite and an active matrix as the presence of rare earth compounds tends to raise hydrogen transfer reactions, reducing the olefin content and consequently the octane number in the cracked naphtha. Another alternative to improve the yield of gasoline in the operation mode is changing the final boiling point of cracked naphtha to higher values. In this case, the limitation is the quality requirements, mainly the sulfur content in the final derivative, according to the operating capacity in the cracked naphtha hydrodesulfurization unit.

The main restrictions in the processing unit in maximum gasoline mode are the gas separation section capacity, especially related to the cold area compressors, and the debutanizer columns.

2. **Maximum LPG:** In this operation mode, the FCC unit operates under high severity translated to high operation temperature (TRX) and high catalyst/oil ratio. The catalyst formulation considers higher catalyst activity through the addition of ZSM-5 zeolite. There is the possibility of a reduction in the total processing capacity due to the limitations in blowers and cold area capacity.

An improvement in the octane number of cracked naphtha despite a lower yield is observed due to the higher aromatic concentration in the cracked naphtha. In some cases, the refiner can use the cracked naphtha recycle to improve the LPG yield even more.

In the maximum LPG operation mode, the main restrictions are the cold area processing capacity, metallurgic limits in the hot section of the unit, treating section processing capacity, and the top systems of the main fractionating column.

3. **Maximum LCO:** The maximum LCO (light cycle oil) operation mode is used by refiners with great demand for middle distillates, especially diesel, and adequate hydrotreating capacity to convert the LCO into high-quality diesel. In this case, the FCC unit operates under relatively low-severity conditions with low TRX and low catalyst/oil ratios, and the catalyst formulation tends to minimize the catalyst activity.

In processing units where a restriction is observed related to the cold area and blower processing capacity, the maximum LCO operation mode can allow the rise in the total processing capacity of the unit. This fact can be positive once it allows lower time contact between catalyst and feed, improving the LCO and decanted oil (DO) yield even more.

It's important to consider the effect of the feed stream quality over the produced LCO. Paraffinic feeds tend to produce higher quality LCO. Refiners operating FCC units in maximum LCO mode tend to minimize the final boiling point of cracked naphtha and maximize this parameter in the LCO aiming to improve the LCO yield, but this action is limited by the quality of the final diesel.

4. **Maximum Aromatic Residue:** This is the less common operation mode in FCC units, where the main objective is to maximize the yield of decanted oil and achieve the quality requirements of aromatic residue. The aromatic residue is normally applied to produce black carbon. This derivative presents a great demand in some markets.

Catalytic Conversion Processes

The main difficulty in complying with the aromatic residue specification regards ash content in the decanted oil. This parameter is strictly related to the cyclone efficiency in the catalyst regeneration section. To achieve this objective, some refiners apply additives to promote ash decantation in the final tanks or specific filtration systems that require more capital spending.

Another key quality parameter to meet aromatic residue specification is the BMCI (Bureau of Mines Correlation Index), which is related to the aromaticity of the decanted oil. To achieve the current specifications of black carbon, it's necessary to achieve a minimum BMCI higher than 120. The BMCI is calculated based on the viscosity of the decanted oil at the temperature of 210°F. The metal content in the decanted oil needs to be also controlled, especially sodium, aluminum, and silicon.

The operating severity in maximum aromatic residue mode tends to be high with high TRX, high catalyst/oil ratio, and high catalyst activity. A side effect is observed, which is the increase of octane number in cracked naphtha due to the incorporation of aromatic compounds in this intermediate.

In maximum aromatic residue operation mode, the main restrictions are the temperature of the bottom section in the main fractionators that can lead to coke formation, metallurgic limitations in the hot sections, as well as the capacity of blowers and cold area compressors.

4.5 THE PETROCHEMICAL FCC ALTERNATIVE: RAISING COMPETITIVE ADVANTAGE

Considering the current scenario of the downstream industry and the last forecasts, the trend of reduction in transportation fuel demand is observed, followed by a growing market of petrochemicals, leading the refiners to optimize their FCC units to maximize LPG yield, aiming to improve the capacity to produce light olefins and promote closer integration with petrochemical assets. A major part of the catalytic cracking units is optimized to maximize transportation fuels, especially gasoline. However, in the face of the current scenario, some units have been optimized to maximize the production of light olefins (ethylene, propylene, and butenes). As aforementioned, units focused on this goal have these operational conditions severely changed, raising the cracking rate.

The reaction temperature reaches 600°C, and a higher catalyst circulation rate raises the gas production, which requires a scaling up of the gas separation section. The higher thermal demand makes operating the catalyst regenerator advantageous in total combustion mode, leading to the necessity of installing a catalyst cooler system, as observed also in the residue FCC units.

The Installation of a catalyst cooler system raises the processing unit profitability through the total conversion enhancement and selectivity to noblest products such as propylene and naphtha against gases and coke production. The catalyst cooler is necessary when the unit is designed to operate under total combustion mode due to the higher heat release rate, as presented here:

$$C + \tfrac{1}{2} O_2 \rightarrow CO \text{ (Partial Combustion)} \quad \Delta H = -27 \text{ kcal/mol}$$
$$C + O_2 \rightarrow CO_2 \text{ (Total Combustion)} \quad \Delta H = -94 \text{ kcal/mol}$$

In this case, the temperature of the regeneration vessel can reach values close to 760°C, leading to higher risks of catalyst damage, which is minimized by installing a catalyst installation. The decision by the total combustion mode needs to consider the refinery thermal balance, once, in this case, will not be possible to produce steam in the CO boiler. Furthermore, the higher temperature in the regenerator requires materials with the noblest metallurgy. This significantly raises the installation costs of these units, which can be prohibitive to some refiners with restricted capital access.

4.5.1 Propylene Production from FCC

Currently, a major part of the propylene market is supplied by steam cracking units, but a great part of the global propylene demand is from the separation of LPG produced in FCC units. Figure 4.5 presents a feedstocks and derivatives profile in a typical FCC unit.

Normally, the LPG produced in FCC units contains close to 30% of propylene, and the added value of the propylene is close to 2,5 times of the LPG. According to the local market, the installation of propylene separation units presents an attractive return on investment. Despite the advantage, a side effect of the propylene separation from LPG is that the fuel stays heavier, leading to specifications issues, mainly in colder regions. In these cases, alternatives are to segregate the butanes and send this stream to the gasoline pool, add propane to the LPG, or add LPG from natural gas. It's important to consider that some of these alternatives reduce the LPG offer, which can be a severe restriction according to the market demand.

A great challenge in the propylene production process is the propane and propylene separation step. The separation is generally hard by simple distillation because the relative volatility between propylene and propane is close to 1,1. This fact generally conducts to distillation columns with many equilibrium stages and high internal reflux flow rates.

There are two technologies normally employed in propylene-propane separation towers that are known as heat-pump and high-pressure configurations.

The high-pressure technology applies a traditional separation process that uses a condenser with cooling water to promote the condensation of top products. In this case, it's necessary to apply sufficient pressure to promote the condensation of

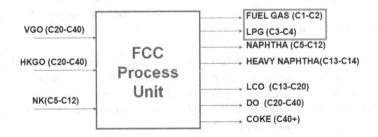

FIGURE 4.5 Production and Feedstocks Profile in a Typical FCC Unit

Catalytic Conversion Processes

products at the ambient temperature. Furthermore, the reboiler uses steam or another available hot source. The adoption of a high-pressure separation route requires a great availability of low-pressure steam in the refining hardware. In some cases, this can be a restrictive characteristic, and the heat pump configuration is more attractive, despite the higher capital requirements.

The separation process applying heat pump technology uses the heat supplied by the condensation of top products in the reboiler. In this case, the reboiler and the condenser are the same equipment. To compensate the non-idealities, it's necessary to install an auxiliary condenser with cooling water.

The application of heat pump technology allows a decrease in the operating pressure by close to 20 bar to 10 bar. This fact increases the relative volatility of propylene-propane, making the separation process easier and, consequently, reducing the number of equilibrium stages and internal reflux flow rate required for the separation.

Normally, when the separation process by distillation is hard (with relative volatilities lower than 1,5), the use of heat pump technology is more attractive.

Furthermore, some variables need to be considered during the choice of the best technology for the propylene separation process, like the availability of utilities, temperature gap in the column, and installation cost.

Normally, propylene is produced in the refineries with specifications. The polymer grade is the most common and has higher added value with a purity of 99,5% (minimum). This grade is directed to the polypropylene market. The chemical grade where the purity varies between 90% and 95% is normally directed to other uses. A complete process flow diagram for a typical propylene separation unit applying heat pump configuration is presented in Figure 4.6.

The LPG from the FCC unit is pumped to a depropanizer column, where the light fraction (essentially a mixture of propane and propylene) is recovered at the top of the column and sent to a deethanizer column, while the bottom (butanes) is pumped to LPG or gasoline pool, according to the refining configuration. The top stream of the deethanizer column (lighter fraction) is sent back to the FCC, where it is incorporated into the refinery fuel gas pool or, in some cases, can be directed to petrochemical plants to recover the light olefins (mainly ethylene) present in the stream, while the bottom of the deethanizer column is pumped to the C_3 splitter column, where the separation of propane and propylene is carried out. The propane recovered in the bottom of the C_3 splitter is sent to the LPG pool, where the propylene is sent to the propylene storage park. The feed stream passes through a caustic wash treatment aiming to remove some contaminants that can lead to deleterious effects on petrochemical processes. An example is carbonyl sulfide (COS), which can be produced in the FCC (through the reaction between CO and S in the riser).

Nowadays, the falling demand for transportation fuel has made the refiners optimize the FCC units, aiming to maximize the propylene yield, following the trend of closer integration with the petrochemical sector. Among the alternatives to maximize the propylene yield in FCC units is the use of ZSM-5 as an additive to the FCC catalyst, as well as the adjustment of the process variables to most severe conditions, including higher temperatures and catalyst circulation rates. Another interesting alternative is to recycle the cracked naphtha to the processing unit aiming to improve the LPG and, consequently, the propylene yield.

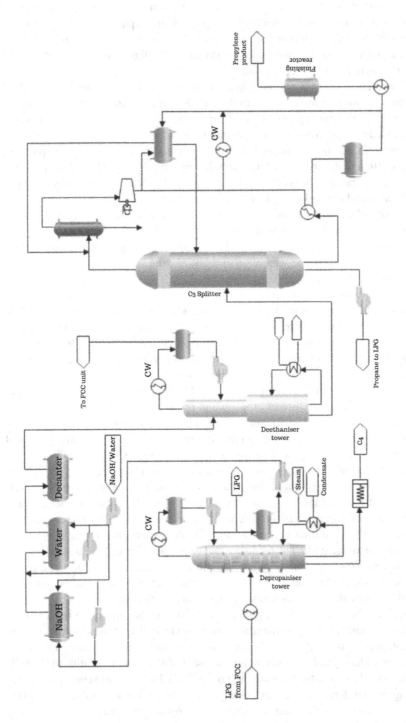

FIGURE 4.6 Typical Process Flow Diagram for an FCC Propylene Separation Unit Applying Heat Pump Configuration

Catalytic Conversion Processes

The installation of propylene separation units can present a significant capital investment to refiners, but considering the last forecasts that reinforce the trend of reduction in the demand for transportation fuels, this investment can be a strategic decision for all players in the downstream industry in the middle term to ensure higher added value to the processed crude oil and market share.

4.6 CATALYTIC REFORMING TECHNOLOGIES

The main objective of the catalytic reforming unit is to produce a stream with high aromatic hydrocarbon content that can be directed to the gasoline pool or to produce petrochemical intermediates (benzene, toluene, and xylene) according to the market served by the refiner. Due to the high content of aromatic compounds, the reformate can significantly raise the octane number in gasoline. In the current scenario, this is a less attractive route.

A typical feedstock to the catalytic reforming unit is the straight-run naphtha. However, in the last decades, due to the necessity to increase the refining margin through the installation of bottom barrel units, hydrotreated coke naphtha stream has been consumed like feedstock in the catalytic reforming unit.

The catalyst generally employed in the catalytic reforming process is based on platinum (Pt) supported by alumina treated with chlorinated compounds to raise the support acidity. This catalyst has bifunctional characteristics once the alumina acid sites are active in molecular restructuring and the metal sites are responsible for hydrogenation and dehydrogenation reactions.

The main chemical reactions involved in the catalytic reforming process are as follows:

- Naphthene compound dehydrogenation
- Paraffin isomerization
- Naphthene compound isomerization
- Paraffin dehydrocyclization

Among the undesired reactions that can be cited are hydrocracking reactions and dealkylation of aromatic compounds.

Figure 4.7 presents a basic process flow diagram for a typical semi-regenerative catalytic reforming unit.

The naphtha feed stream is blended with recycled hydrogen and heated at a temperature varying from 500°C to 550°C before entering in the first reactor. As the reactions are strongly endothermic, the temperature falls quickly, so the mixture is heated and sent to the second reactor, and so on. The effluent from the last reactor is sent to a separation drum, where the liquid and gaseous phases are separated.

The gaseous stream with high hydrogen content is shared in two process streams. A part is recycled to the process to keep the ratio H_2/feed stream; the other part is sent to a gas purification process plant (normally a pressure swing adsorption unit) to raise the purity of the hydrogen that will be exported to other process plants in the refinery.

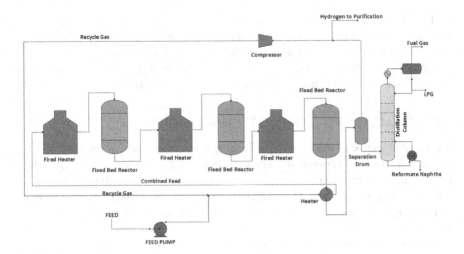

FIGURE 4.7 Typical Arrangement to Semi-regenerative Catalytic Reforming Processing Unit

The liquid fraction obtained in the separation drum is pumped to a distillation column wherein the bottom is produced the reformate, and in the top drum of the column, LPG and fuel gas are produced.

The reformate has a high aromatic content and, according to the market supplied by the refinery, can be directed to the gasoline pool like a booster of octane number or when the refinery has aromatic extraction plants is possible to produce benzene, toluene, and xylene in segregated streams, which can be directed to petrochemical process plants. The gas rich in hydrogen produced in the catalytic reforming unit is an important utility for the refinery, mainly when there is a deficit between the hydrogen production capacity and the hydrotreating installed capacity in the refinery. In some cases, the catalytic reforming unit is operated with the principal objective of producing hydrogen.

One of the main process variables in the catalytic reforming processing unit is pressure (3,5–30 bar), which is normally defined in the design step. In other words, pressure is normally not an operational variable. The temperature can vary from 500°C to 550°C. The space velocity can be varied through feed stream flow rate control and the ratio H_2/feed stream that have a direct relation with the quantity of coke deposited on the catalyst during the process. In semi-regenerative units, the ratio H_2/feed stream can vary from 8 to 10. In units with continuous catalyst regeneration, this variable can be significantly reduced.

Due to the process severity, the high coke deposition rate on the catalyst and consequently the quick deactivation lead to short operational campaign periods and semi-regenerative units that employ fixed-bed reactors.

To solve this problem, some technology licensors developed a catalytic reforming process with continuous catalyst regeneration steps.

The process Aromizing, developed by Axens, uses side-by-side configurations to the reactors, while CCR Platforming, developed by UOP, uses the configuration of

Catalytic Conversion Processes

stacked reactors for the catalytic reforming process with continuous catalyst regeneration. Figure 4.8 presents a flow diagram of the Aromizing catalytic reforming unit.

Both technologies are commercial, and some process plants with these technologies are in operation around the world. Figure 4.9 presents a basic process flow diagram for CCR Platforming developed by UOP.

In the regeneration section, the catalyst is submitted to processes to burn the coke deposited during the reactions and treated with chlorinated compounds to reactivate the acid function of the catalyst.

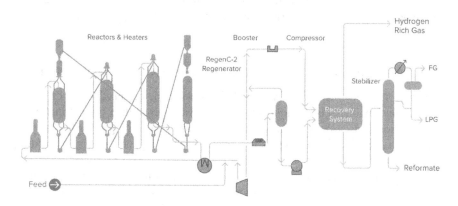

FIGURE 4.8 Aromizing Reforming Technology by Axens (with Permission)

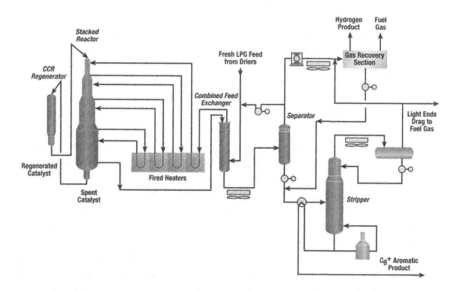

FIGURE 4.9 CCR Platforming Reforming Technology by UOP (with Permission)
Source: Honeywell UOP (www.uop.honeywell.com)

Despite the higher capital investment, the rise in the operational campaign and higher flexibility in relation to the feedstock to be processed in the processing unit can compensate for the higher investment in relation to the semi-regenerative process.

The catalytic reforming technology gives great flexibility to the refiners in the gasoline production process. However, in the last decades, there has been a strong restriction on the use of reformate in the gasoline due to the control of benzene content in this derivate (due to the carcinogenic characteristics of this compound). This fact has reduced the application of reformate in the gasoline formulation in some countries. Furthermore, the operational costs are high, mainly due to the catalyst replacement and additional security requirements linked to minimizing leaks of aromatic compounds.

4.6.1 Aromatics Separation Section: Ensuring Maximum Added Value

As aforementioned, in markets where there is demand, the production of petrochemical intermediates is economically more advantageous than the production of transportation fuels, especially in countries with easy access to lighter oils. The production and separation of aromatics are processes with a great capacity for adding value to crude oil.

The aromatic production complex is a set of processes intended to produce petrochemical intermediates from naphtha produced in the catalytic reforming process or by the pyrolysis process. An aromatic production complex can take on different process configurations according to the petrochemical market to be served. An example is shown in Figure 4.10.

The naphtha rich in aromatics, produced in catalytic reforming or pyrolysis units (in some cases from both), is fed to an extractive distillation column where the separation of aromatic compounds is conducted, which are drawn in the extract phase and are recovered at the bottom of the column, while the non-aromatic compounds are drawn from the top in the raffinate phase. The aromatics are separated from the solvent in the solvent recovery column and directed to the fractionation section of aromatics, where the essentially pure benzene and toluene streams and xylenes blend are obtained. The raffinate is sent to a wash column, and the non-aromatic hydrocarbons are usually sent to the refinery's gasoline pool.

The process shown in Figure 4.10 involves only physical separation steps; that is, the process yields in a given stream depend on the concentration of this compound in the feed stream.

The growing demand for high-quality petrochemical intermediates and the higher added value of these products have made it necessary to develop conversion processes capable of converting lower interest aromatics (toluene) into more economically attractive compounds (xylene).

The separation of aromatics, mainly xylenes, is a great challenge to modern engineering. The similarities between the molecules make the separation through simple distillation very hard. For this reason, several researchers and technology licensors dedicate their efforts to developing new processes that can lead to pure

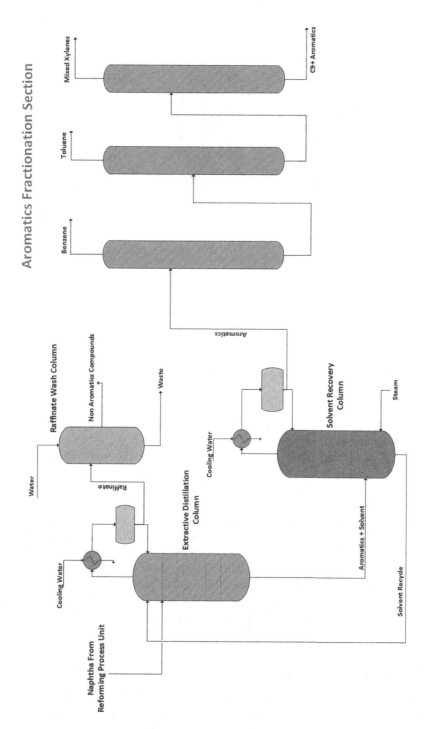

FIGURE 4.10 Basic Process Configuration for a Typical Aromatics Separation Unit

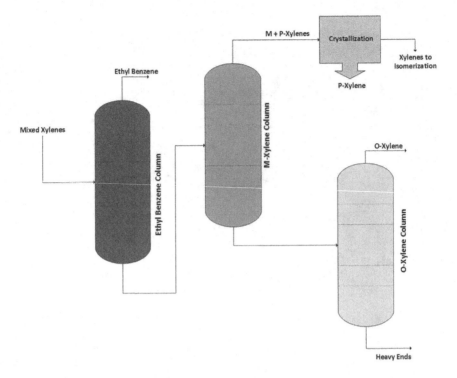

FIGURE 4.11 Basic Process for Xylene Separation

compounds with lower costs. A basic scheme for a xylene separation process is shown in Figure 4.11.

The xylene blend is fed to a distillation column where the ethylbenzene is separated in the top and sent to the styrene production market, while the bottom stream is pumped to another column where the mixture of meta- and para-xylenes is drawn from the top and the ortho-xylene and heavier compounds are removed in the bottom.

Ortho-xylene is separated from heavy aromatics in another distillation column, while the meta- and para-xylenes are fed to a crystallization process, where a stream is obtained with a high concentration in meta-xylene and the residual stream is directed to an isomerization unit, aiming to promote the conversion of residual meta- and ortho-xylenes in para-xylene. The aromatic production units are normally optimized to maximize the para-xylene production because this is a petrochemical intermediate with higher interest. This compound is raw material to produce terephthalic acid that is used to produce PET (polyethylene terephthalate).

4.6.2 Improving the Yield of Light Aromatics: Molecular Management

To raise the production of higher commercial and economic interest compounds (p-xylene and benzene), technology licensors developed several processes to convert streams with low added value in these compounds. One of the main developers of this technology is UOP. The Parex process uses the separation through adsorption to obtain high purity p-xylene from the xylene blend.

Another UOP technology is the Isomar process, which promotes the xylene isomerization to para-xylene, raising the recovery of this compound in the aromatic complex. The Tatoray process was developed to convert toluene and heavy aromatics (C_{9+}) into benzene and xylene through the transalkylation reaction. Another economically attractive technology is the Sulpholane process, which applies liquid-liquid extraction operations and extractive distillation to reach high purity aromatics separation from hydrocarbon mixture.

UOP developed an integrated aromatics complex aiming to maximize the production of benzene and p-xylene, which leads to higher profitability for the refiner. A UOP aromatics complex scheme is presented in Figure 4.12.

Other companies have attractive and efficient technologies to produce high purity aromatics. The Axens licenses an aromatic production complex also based on separation and conversion processes, called ParamaX, which can be optimized to produce p-xylene. The ParamaX technology offers the possibility of cyclohexane production (raw material to synthetic fibers) through benzene hydrogenation beyond raising the production of this component through toluene hydrodealkylation (I).

As aforementioned, the capital investment in the installation of aromatic production complexes is high. However, the obtained products have high added value and rely on great demand, and even the compounds with low interest can be commercialized with a high margin. In countries with easy access to light oil reserves, such as Saudi Arabia and the United States (tight oil), the installation of these process plants is even more economically attractive.

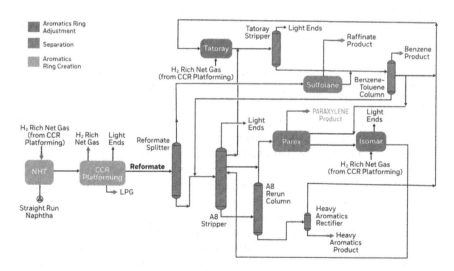

FIGURE 4.12 Aromatics Complex by UOP (with Permission)

Source: Honeywell UOP (www.uop.honeywell.com)

4.7 NAPHTHA ALKYLATION TECHNOLOGIES

Technological development associated with restrictive environmental legislation has led to the development of internal combustion engines, which need higher performance fuels, lower consumption, and low emission of environmentally harmful gases (NOx, SOx, etc.).

This scenario presents a growing challenge to the crude oil refining industry and process technology developers, mainly when the maturation of new transport alternatives like electric vehicles and hybrid engines strongly threatens the market for crude oil derivates, such as diesel and gasoline.

Gasoline is one of the most consumed crude oil derivates and is normally produced through a mixture of naphtha from different refining process steps. The streams normally involved in the gasoline production process are straight-run naphtha, cracked naphtha, coke naphtha (after hydrotreatment), and reforming naphtha.

One of the main parameters of gasoline quality is the octane number, which is a measure of the combustion quality of this derivate. One of the streams that contribute to raising the octane number is reformed naphtha, produced in the catalytic reforming unit. However, due to the severe restrictions related to the carcinogenic aromatic emissions, mainly benzene, some refiners have avoided the application of this stream to formulate gasoline, directing the reformed naphtha preferably to petrochemical intermediate production in aromatic complexes.

An alternative to the reforming naphtha is the production of branched hydrocarbons (with a high octane number) through the catalytic alkylation process.

The alkylation process involves the reactions between light olefins (C_3–C_5) and isoparaffinic hydrocarbons like isobutane. The reaction product called alkylate is a mixture of branched hydrocarbons with higher molecular weight and higher octane number.

An example of a typical alkylation reaction is represented here:

$$C_4H_{10} + C_3H_6 \rightarrow C_7H_{16} \text{ (2,3 Dimethylpentane)}$$

The reaction is catalyzed in a strongly acidic reaction environment. The acids normally employed in industrial-scale technologies are hydrofluoric acid (HF) and sulfuric acid (H_2SO_4).

The main advantage of the alkylation process is the production of a stream with a high octane number, high chemical stability, and freedom from contaminants such as nitrogen and sulfur. These characteristics turn the alkylate a component attractive to the gasoline formulation in the automotive and aviation industries.

Alkylation process feed streams are generally obtained from LPG produced in deep conversion units, mainly FCC and delayed coking. The LPG produced in these processing units has high olefin content, ideal for the alkylation process. The isobutane stream is normally obtained through the separation of LPG produced in the atmospheric distillation unit, FCC, or delayed coking in deisobutanizer towers.

As aforementioned, the acids generally employed as the homogeneous catalyst for the alkylation process are HF and H_2SO_4. Figure 4.13 presents a process flow diagram of the alkylation process catalyzed by HF.

Catalytic Conversion Processes

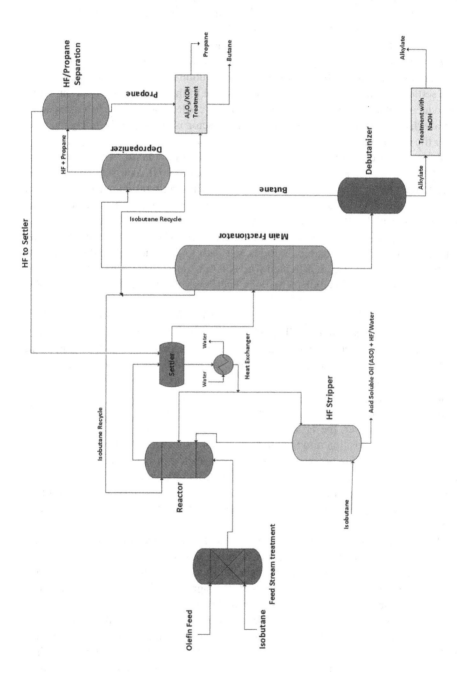

FIGURE 4.13 Typical Process Flow Diagram to Catalytic Alkylation Unit Using HF as Catalyst

The feed stream goes through a pretreatment (generally molecular sieves or alumina) before being pumped to the reactor. The objective is to remove process contaminants, mainly water, diolefins, and sulfur and nitrogen compounds. Water is especially damaging to the process as it accelerates piping and equipment corrosion process, and beyond requires higher HF reposition.

After pretreatment, the hydrocarbon streams are put in contact with the HF in the reactor, and the hydrocarbon mixture and HF solution are separated through gravity in a settler vessel. The hydrocarbon phase is sent to the fractionating section, while the aqueous phase (containing the most of HF) is cooled and sent back to the reactor. As alkylation reactions are exothermic, the reactor is continuously refrigerated to keep the reaction in ideal conditions.

A part of hydrofluoric acid is sent to the stripping column, where the acid is stripped with isobutane. The top product is a mixture of HF and isobutene and sent back to the reactor, while the bottom stream contains an azeotropic mixture of water and HF. Beyond hydrocarbons, this step is responsible for keeping the HF free of contaminants and with adequate concentration for the alkylation process.

After the separation columns, butane and propane streams go to treatment with alumina to decompose organic fluorides and with KOH to neutralize the remaining acidity. The alkylate stream is treated with NaOH to neutralize the remaining acidity. Currently, the main alkylation technology licensors with HF are the companies UOP and ConocoPhillips.

Alkylate stream is normally directed to the refinery gasoline pool for the production of high-octane automobile gasoline or aviation gasoline. However, in petrochemical plants, this stream can be used as an intermediate to produce ethylbenzene (for styrene production), isopropyl-benzene (to produce phenol and acetone), and dodecyl-benzene (used to produce detergents). Propane and butane streams can be sent to the LPG pool of the refinery or commercialized separately.

The alkylation process with sulfuric acid as a catalyst has similarities with the HF process. However, the H_2SO_4 regeneration step is more complex and involves the H_2SO_4 decomposition into SO_2 and SO_3 and the subsequent condensation of concentrated H_2SO_4. This regeneration can be conducted in the processing site or in an external process plant. Consequently, the H_2SO_4 consumption in the process is much higher than that of HF. Furthermore, the solubility of H_2SO_4 in hydrocarbons is lower, requiring greater agitation to maintain the contact between the phases in ideal conditions for the process.

The alkylation technologies with sulfuric acid most applied on an industrial scale are the processes Effluent Refrigerated Alkylation Process (licensed by Stratco Engineering) and Cascade AutorefrigeratedProcess (licensed by ExxonMobil). Figure 4.14 shows a simplified process flow diagram of the alkylation technology with H_2SO_4, licensed by Stratco Engineering.

Olefin feed streams go to a coalescer to remove water. After the mixture with isobutane recycled, the mixture is sent to the reactor in the sequence. The mixture of hydrocarbons and acid follows to a settler where the phase separations occur. The organic phase is sent back to the reactor. A control valve promotes the necessary pressure reduction to vaporize the lighter hydrocarbons and remove heat from the reactor, controlling the equipment temperature, which rises due to the exothermic characteristics of alkylation reactions.

Catalytic Conversion Processes

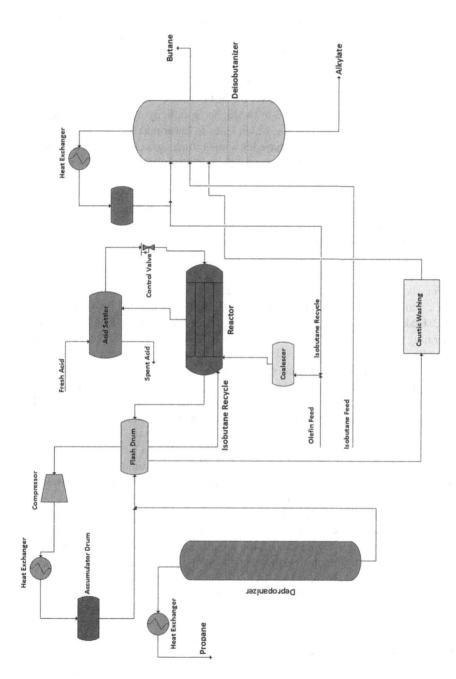

FIGURE 4.14 Basic Process Flow Diagram to the Alkylation Process Catalyzed by H_2SO_4, Developed by the Stratco Engineering Company

The hydrocarbon blend is sent to a flash drum where the lighter phase is directed to a compressor to condense in an accumulator vessel, and the propane is recovered in the depropanizer tower, while the heavier hydrocarbons (essentially isobutane) are recycled to the reactor. The stream containing the alkylate is directed to a caustic treatment and posteriorly to a deisobutanizer column, where the alkylate is removed at the bottom.

As aforementioned, the need for catalyst replacement is higher in the process with sulfuric acid. However, the HF process needs a higher isobutane/olefin ratio, which means a greater separation system. Over the last decades, the refiners have opted for the HF alkylation technologies due to the higher simplicity of this process and the lower need for catalyst replacement, which leads to lower operational costs.

However, regulatory pressures have led some refiners to convert their HF alkylation units to operate with H_2SO_4. Due to the high volatility and higher risks presented by HF, some licensors developed technologies to convert HF units to operate with H_2SO_4 like the Alkysafe technology, licensed by Stratco Engineering Company, and the ReVap process, developed by ExxonMobil and ConocoPhillips, which uses additives to reduce the HF volatility, making the unit operation safer.

The principal process variables of the alkylation process are the isobutane/olefin ratio, reaction temperature, acid/hydrocarbon ratio, acid purity, residence time in the settler, and operation pressure.

For alkylation processes with HF, the reaction temperature varies from 20°C to 40°C, while H_2SO_4 processes operate under lower temperatures, between 4°C and 10°C. Under higher temperatures, H_2SO_4 can undergo decomposition to SO_2 and SO_3. Operating pressure is generally sufficient to keep the hydrocarbons in the liquid phase, normally lower than 5 bar. The residence time in the settler is important because if this time is quite low, undesirable reactions can occur as organic fluoride production, in the case of HF, which raises the catalyst consumption and reduces the alkylate production.

Acid purity must be maintained higher as possible through the removal of ASO (acid-soluble oil), water, and dissolved reactants in the HF case and through fresh acid replacement in processes with acid sulfuric as the catalyst.

The main disadvantage of the alkylation processes with a homogeneous catalyst (HF or H_2SO_4) is the need to handle strong, highly concentrated acid, which leaves greater process safety risks and high maintenance costs, mainly related to avoiding corrosion in piping and equipment.

Aiming to eliminate these risks, some licensors have dedicated their efforts to developing heterogeneous catalysts that can replace the strong acids in the alkylate production processes. UOP developed the process called Alkylene, which applies a solid catalyst with continuous regeneration during the process. Among other technologies that can be cited are the processes Lurgi Eurofuel (developed by the Lurgi Engineering Company in cooperation with Sud-Chemie), AlkyClean (developed by the companies ABB-Lummus and Akzo Nobel), and FBA (developed by Haldor Topsoe).

4.8 NAPHTHA ISOMERIZATION

Reformed naphtha, produced in the catalytic reforming unit, is one of the streams that contribute to raising the octane number in the final gasoline. However, due to the severe restrictions related to the carcinogenic aromatic emissions, mainly benzene,

Catalytic Conversion Processes

some refiners have avoided the application of this stream to formulate gasoline, directing the reformed naphtha preferably to petrochemical intermediate production in aromatic complexes.

An alternative to the reforming naphtha is the production of branched hydrocarbons (with a high octane number) through processes such as catalytic alkylation and isomerization.

The isomerization process involves the conversion of normal paraffin to branched paraffin, keeping the carbon number in the reactant molecule. The reactions are carried out in the presence of hydrogen under pressure and temperature mild conditions. The presence of hydrogen aims mainly to avoid coke deposition on the catalyst.

Isomerization reactions (4.1) are slightly exothermic; that is, they are favored under lower temperatures. To balance the kinetic requirements and thermodynamic limits, high activity catalysts are applied, which allow operation under lower temperatures.

$$N\text{-paraffin} \longleftrightarrow I\text{-paraffin} \qquad (4.1)$$

The catalysts applied in the isomerization processes have bifunctional characteristics containing acids and metallic sites. The most employed catalysts are platinum-impregnated chlorinated alumina and zeolitic and oxide-based catalysts.

The isomerization process produces a light stream that can contribute to raising the octane number to the final gasoline and is practically free of contaminants like nitrogen and sulfur. However, due to the high content of lighter compounds, the isomerate stream negatively affects the gasoline fugitive emissions specification (high Reid vapor pressure); that is, the isomerate needs to be blended with heavier streams to formulate gasoline that meets the current specifications.

The typical feed stream to isomerization units is normally the lighter fraction (C_5–C_6) of straight-run naphtha when the objective is to produce isomerate naphtha, which will be directed to the gasoline pool. Another process that has great interest to the refining industry is the isomerization of N-butane to isobutane. This product can be directed to the feed stream to catalytic alkylation units or to produce MTBE (methyl tert-butyl ether). In this case, the process feed stream is the heavier fraction of LPG obtained in the debutanizer columns. The isomerization catalyst is sensitive to contaminants such as nitrogen and sulfur, and then normally, the feed passes through a treating step to reduce the contaminant content (normally hydrotreatment).

Figure 4.15 shows a process flow diagram for a typical isomerization unit.

The feed stream is mixed with hydrogen and is preheated in heat exchangers, and then it is fed to the reactor, and the reactor effluent is cooled through heat exchange with the fresh feed, and then it is directed to a separation vessel where a major part of hydrogen is separated from the liquid phase. Hydrogen is sent back to the process by the compressor, while the liquid phase is sent to a distillation column where the isomerate naphtha is the bottom product, and the light products are removed from the top.

The main process variables of the isomerization processes are temperature, operation pressure, and space velocity. As aforementioned, temperature is normally reduced and can vary between 120 and 250°C depending on the activity of the employed catalyst due to the isomerization reaction characteristics, which seek to operate under lower temperatures.

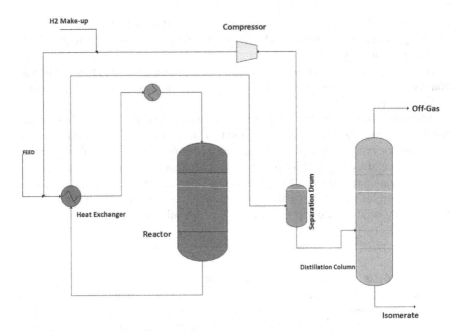

FIGURE 4.15 Typical Arrangement for Single Reaction Stage Isomerization Unit

Operating pressure is normally close to 30 bar. Higher pressures conduct the higher catalyst life cycle and favor hydrocracking reactions that are undesirable. The space velocity in the reactor is a design variable and expresses a relation between residence time and catalyst total cost. Lower velocity results in higher catalyst mass and higher residence time, which allow operation under lower temperature level.

One of the most known isomerization technologies is the Penex process, developed by UOP, which uses platinum-impregnated chlorinated alumina as the catalyst. Figure 4.16 presents a simplified scheme for this process.

Penex technology uses dryer vessels containing molecular sieves to remove water from feed and hydrogen streams, preserving the catalyst.

The process applies two reactors in series, and this arrangement allows the operation of one reactor while the other is under catalyst replacement or maintenance.

Another highly employed isomerization technology is the Par-Isom process, also developed by UOP. In this case, a sulfated-zirconia catalyst is applied with the possibility of regeneration in the processing unit.

GTC Technology developed the Isomalk-2 technology, which also applies platinum-impregnated sulfated zirconia as the catalyst. The process still uses the pre-fractionating step and the recycling of poor octane index streams.

Despite the need for higher investment, the Isomalk-2 technology can produce isomerate naphtha with higher octane index.

As the isomerization reactions are limited by equilibrium conversion, some technologies involve removing isoparaffin formed during the process from the

Catalytic Conversion Processes

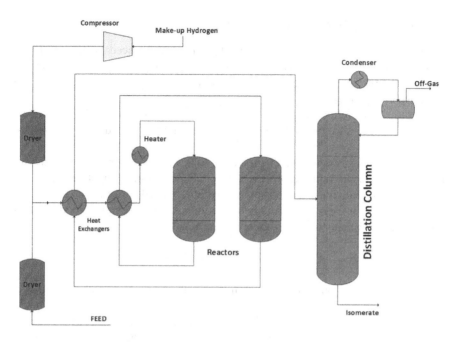

FIGURE 4.16 Simplified Process Scheme to Penex technology by UOP (with Permission)
Source: Honeywell UOP (www.uop.honeywell.com)

recycle stream, shifting the reactional equilibrium, and raising the quality of the final product. These technologies apply sophisticated separation processes like simulated moving beds and molecular sieves. The technologies Ipsorb and Hexsorb (developed by Axens) and Molex and IsoSiv (developed by UOP) are examples of these processes.

As mentioned previously, the N-butane isomerization is economically attractive once the isobutane is applied like feedstock to catalytic alkylation processes, which is another process capable of producing high-quality gasoline. Another use of isobutane is MTBE production. However, with the current environmental regulations, the use of a gasoline additive is falling and, in some countries, is prohibited.

The main butane isomerization technology is the Butamer process, developed by UOP. Another available technology is the Iso-C_4, developed by Axens. The process is basically the same applied to the C_5–C_6 fraction isomerization process as described previously.

Isomerization processes show advantages in relation to other gasoline-upgrading technologies because they can produce a low-contaminant stream (nitrogen and sulfur) without aromatic compounds, and they are productive processes safer when compared with catalytic reforming and alkylation technologies.

These characteristics make the isomerization processes attractive to refiners inserted in markets with high demand for gasoline and petrochemical intermediates.

4.9 LIGHT OLEFIN CONDENSATION

The evolution way of our society has created the growing need for transportation fuels, the economic and technological development of nations can be almost instantly translated by the rise in oil consumption, and much of this consumption is destined for transportation fuels.

Over the years, refiners and researchers have studied strategies and technological routes that aim to raise the high-quality derivative yield, mainly diesel, jet fuel, and gasoline.

In the case of gasoline, one of the oldest routes applied to raising the yield of this derivative in the refining process is the polymerization of light olefins or the so-called condensation of these compounds (propylene and butylene).

The process consists of an olefin combination and the consequent production of high-molecular-weight olefins, as presented as the following:

$$n\ CH_2=CH_2 \rightarrow H\text{-}(CH_2CH_2)_n\text{-}H \qquad (4.2)$$

This process is controlled to maintain the products in the gasoline distillation range.

The reactions are carried out under mild temperature conditions (15–25°C) and pressures that vary between 10 and 80 bar, according to the process feed stream. The condensation reactions are exothermic, and the reaction is controlled by heat removed from the reactor, normally through cold stream injection between the catalytic beds (quench streams). The process can be conducted under thermal or catalytic conditions. However, the commercial units normally apply the catalytic route using acid catalyst, mainly phosphoric acid deposited on an adequate carrier.

Raw material (light olefins) for the process is normally from FCC units, catalytic dehydrogenation, or thermal cracking units.

A typical process flow diagram for a gasoline production unit through olefin condensation is presented in Figure 4.17.

The feed stream passes through a process to remove contaminants, mainly sulfur compounds that are a catalyst poison. From the feed drum, the olefinic stream is pumped to the reactor, where temperature control is realized by the injection of reactants under a reduced temperature between the catalytic beds, as aforementioned.

After the reaction step, the reactor effluent is separated in a flash distillation column, where the bottom product is sent to a stabilizer column that aims to remove the light compounds from the naphtha stream, and the top product is recycled to the process. Normally the gasoline produced by this route is called polymerization gasoline.

This process is capable of producing high-octane gasoline. However, the high olefin content makes the polymerization of gasoline chemically unstable, leading to gum production in the short term. The polymerization of gasoline is normally directed to the refinery gasoline pool and mixed with the other naphtha streams to contribute to raising the octane index of the final gasoline. However, adding is limited too by high volatility (high Reid vapor pressure). The remaining olefins stream can be directed to the petrochemical intermediate market or to the refinery LPG pool.

Catalytic Conversion Processes

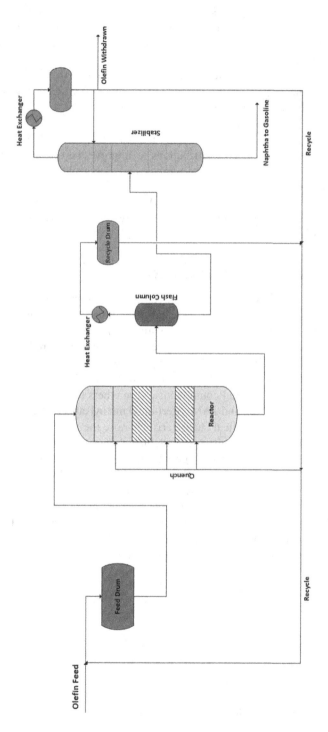

FIGURE 4.17 Typical Process Flow Diagram for Light Olefin Condensation Process to Produce Gasoline

The gasoline production process by olefins condensation or polymerization lost space to other technologies like isomerization, alkylation, and catalytic reforming that can produce more chemically stable naphtha with the same octane index (in some cases higher).

Nowadays, light olefins are intended for consumption as petrochemical intermediates due to the higher profitability offered to the refiner. However, refiners inserted in markets with high gasoline demand can apply this process to quickly raise the high-quality gasoline production.

Some important technology developers such as UOP, Shell Global Solutions, Lummus, and Axens developed contributions to improve the olefins condensation process. The OCT technology from Lummus and Polynaphtha from Axens are among the commercial technologies.

Although currently obsolete, as quoted earlier, polymerization or olefin condensation can contribute an agile and relatively cheap way to raise gasoline production, mainly in extreme scenarios of the shortage of derivatives.

4.10 ETHERIFICATION TECHNOLOGIES

Over the last decades, the development of more efficient engines associated with higher consumption and severe environmental regulations create the need for better gasoline additives development, mainly in relation to the environmental impact produced by these compounds. Previously applied, lead additives were banned due to the strong toxicity and environmental impact produced by these chemicals.

Among the additives that were widely applied in the gasoline formulation, it's possible to highlight oxygenated compounds as the ethers MTBE (tert-butyl ether), ETBE (ethyl tert-butyl ether), and TAME (tert-amyl methyl ether). These compounds, due to their physical and chemical characteristics, raise the gasoline octane number and, in shortage scenarios, can raise the available volume of this crude oil derivative.

These additives are produced through etherification reactions, which consist of the addition of alcohol to olefin-producing ethers. MTBE and TAME are produced from methanol, while the ETBE is produced from ethanol.

Due to their characteristics (total miscibility with hydrocarbons, high octane number, and low volatility), MTBE is the ether most widely employed. The MTBE production chemical reaction is presented in Figure 4.18.

The reaction is catalyzed by an acid catalyst with cationic resins as the carrier.

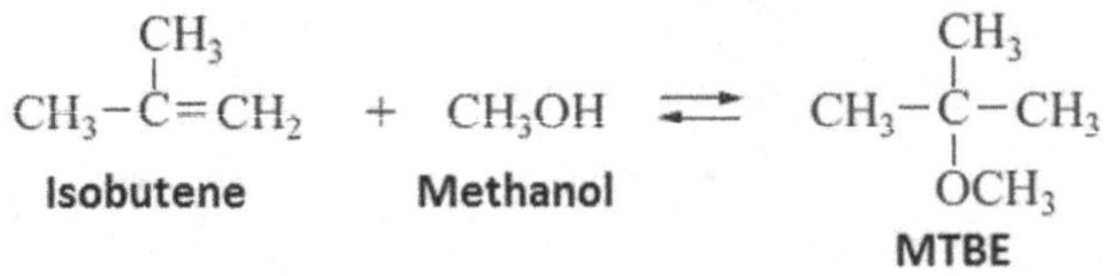

$$CH_3-\underset{\underset{\text{Isobutene}}{|}}{\overset{\overset{CH_3}{|}}{C}}=CH_2 + \underset{\text{Methanol}}{CH_3OH} \rightleftharpoons CH_3-\underset{\underset{\underset{\text{MTBE}}{OCH_3}}{|}}{\overset{\overset{CH_3}{|}}{C}}-CH_3$$

FIGURE 4.18 MTBE Production Chemical Reaction

Catalytic Conversion Processes

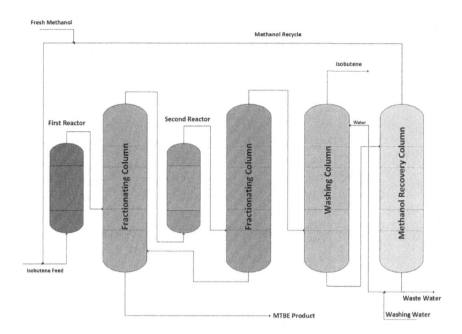

FIGURE 4.19 Typical Process Arrangement to MTBE Production Processing Unit

In refineries with MTBE-producing units, the isobutene is normally obtained through fractionating of LPG from FCC units. MTBE production reaction is exothermic and is carried out under mild conditions. Normally the pressure varies from 10 to 20 bar, and the temperature is controlled from 40°C to 70°C. Figure 4.19 shows a typical process flow scheme for MTBE production.

The fresh methanol feed is mixed with the recycled methanol and posteriorly mixed with the isobutene stream before entering the first reactor. The first reactor effluent is sent to a fractionating tower where the bottom product is the MTBE, and the top stream is directed to the second reaction stage. The effluent of this reactor is pumped to another distillation column where the MTBE is removed from the bottom, and the top is directed to a water wash column. In this column, the remaining isobutene is removed at the top, and the methanol/water mixture is removed in the bottom. This mixture is separated in another fractionating column, and the remaining methanol is recycled to the process while the water is removed for treatment or partially recycled to the washing tower.

The principal process variables for the MTBE production process are pressure and temperature, which need to be controlled at the lower possible level to attend to the compromise between the kinetic requirements and the exothermic characteristics of the reaction.

Main process technology licensors developed technologies to produce oxygenated additives for the gasoline, mainly MTBE.

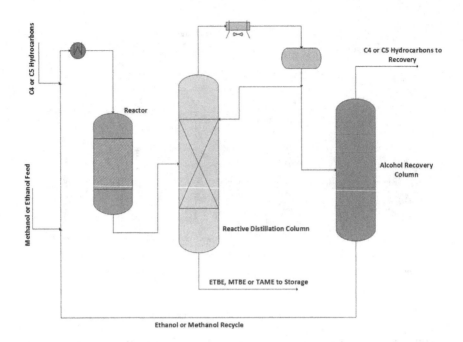

FIGURE 4.20 Simplified Process Flow Diagram for Ethermax Technology by UOP (with Permission)

Source: Honeywell UOP (www.uop.honeywell.com)

One of the main commercialized technologies is the Ethermax process developed by UOP; this process is presented in Figure 4.20.

This process is capable of producing MTBE, ETBE, and TAME, according to the raw material applied. This fact gives great flexibility to the process, which is a considerable advantage. A major part of the etherification reactions occurs in the reactor, and the remaining alcohol is partially converted in the reactive distillation column.

As aforementioned, some of the main technology licensors developed processes aiming to produce oxygenated additives, mainly in the '80s and '90s, with the objective to maximize the production of these compounds in substitution of lead-based additives. Among the main commercially available technologies, we can cite the CDMTBE process, developed by Lummus, and the CATACOL technology, developed by Axens.

Although these compounds have been widely used, in the last few years, ethers as a gasoline additive have been reduced. Only a few countries still use these compounds in gasoline. In the '90s, some cases of water contamination with MTBE in the USA led this country to ban the use of this compound. This fact reduced the available market severely. After this decision, some refiners deactivated their MTBE production plants.

In Brazil, the use of ethers as a gasoline additive was banned in 1996. In our case, the decision is motivated by environmental and socioeconomic reasons. Brazil is a great ethanol producer, and the use of this compound as a gasoline additive (20–27% in volume) is a way to maintain the market to of producers. Beyond this, ethanol is widely used as a transportation fuel in Brazil. This characteristic helps to reduce the pressure upon the internal gasoline demand.

As cited earlier, relatively recent studies proved the toxicity of ethers, which led the refiners to substitute these compounds. Nowadays, the gasoline octane number is normally raised through ethanol addition or raising the percentage of naphtha from processes like catalytic reforming, isomerization, and catalytic alkylation. However, in the technological point of view, the etherification processes are still very interesting.

4.11 LIGHT PARAFFIN DEHYDROGENATION

Light paraffin is normally commercialized as LPG or gasoline and presents reduced added value when compared with light olefins.

The dehydrogenation process involves hydrogen removed from the paraffinic molecule and consequently hydrogen production, according to the reaction (4.3):

$$R_2CH\text{-}CHR_2 \leftrightarrow R_2C=CR_2 + H_2 \qquad (4.3)$$

The dehydrogenation reactions have strongly endothermic characteristics, and the reaction conditions include high temperatures (close to 600°C) and mild operating pressures (close to 5 bar). The catalyst normally applied in the dehydrogenation reactions is based on platinum carried on alumina (other active metals can be applied).

Figure 4.21 shows a schematic process flow diagram for a typical dehydrogenation processing unit.

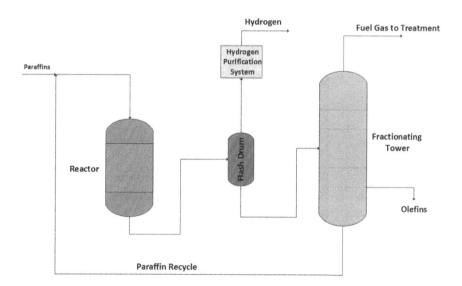

FIGURE 4.21 Process Flow Diagram for a Typical Light Paraffin Dehydrogenation Processing Unit

The main processes that can produce streams rich in light paraffin are physical separation processes such as LPG from atmospheric distillation and units dedicated to separating gases from crude oil.

The feed stream is mixed with the recycle stream before to entre to the reactor, the products are separated into fractionating columns, and the produced hydrogen is sent to purification units (normally PSA units) and posteriorly sent to consumers units as hydrotreating and hydrocracking, according to refining scheme adopted by the refiner. Light compounds are directed to the refinery or petrochemical complex fuel gas pool after adequate treatment, while the olefinic stream is directed to the petrochemical intermediate consumer market.

During the dehydrogenation process, there is a strong tendency for coke deposition on the catalyst surface, and the regeneration of the catalytic bed is periodically carried out through controlled combustion of the produced coke. Some process arrangements present two reactors in parallel aim to optimize the processing unit's operational availability. In these cases, while a reactor is in production, the other is in the regeneration step.

UOP developed and commercialized the Oleflex, which is capable of producing olefins from paraffin dehydrogenation with a continuous catalyst regeneration process. Despite the higher initial investment, this technology can minimize the unavailability period to regenerate the catalyst.

Another available technology is the Catofin process, licensed by Lummus, as aforementioned. In this case, two reactors are used in parallel, as presented in Figure 4.22.

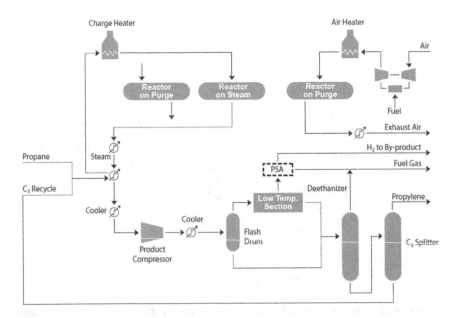

FIGURE 4.22 Simplified Process Scheme to Catofin Dehydrogenation Technology, by Lummus (with Permission)

Other dehydrogenation technologies available are the processes Star, by ThyssenKrupp-Uhde, and FBD, by SnamProgetti Company.

Due to their chemical characteristics, olefinic compounds can be employed in the production of a large number of interesting products, such as polymers (polyethylene and polypropylene), propylene oxide, and oxygenated compound production intermediates (MTBE, ETBE, etc.).

As a process of high energy consumption, there is a great variety of research in the sense of developing more active and selective catalysts that reduce the need for energetic contribution to the dehydrogenation process. One of the main variations of the dehydrogenation process is the process called oxidative dehydrogenation, which occurs according to Reaction 4.4.

$$R_2CH\text{-}CHR_2 + O_2 \leftrightarrow R_2C\text{=}CR_2 + H_2O \tag{4.4}$$

This reaction is strongly exothermic, and this is the main advantage in relation to the traditional dehydrogenation process. On the other hand, the main problem is the strong probability of paraffin combustion in the dehydrogenation reaction.

Refiners inserted in a market with high demand for petrochemical intermediates can find that dehydrogenation technologies are strong allies in the sense that they minimize the yield in reduced-added-value products and raise profitability. In scenarios increasingly competition, these alternative routes acquire strategic character perennially in the crude oil downstream industry.

BIBLIOGRAPHY

1. Fahim, M.A., Al-Sahhaf, T.A., Elkilani, A.S. *Fundamentals of Petroleum Refining*. 1st edition, Elsevier Press, 2010.
2. Myers, R.A. *Handbook of Petroleum Refining Processes*. 3rd edition, McGraw-Hill, 2004.
3. Robinson, P.R., Hsu, C.S. *Handbook of Petroleum Technology*. 1st edition, Springer, 2017.
4. The Catalyst Group. *Advances in Catalysis for Plastic Conversion to Hydrocarbons*, The Catalyst Group (TCGR), 2021.
5. Albahar, M.Z. *Selective Toluene Disproportionation over ZSM-5 Zeolite*, PhD Thesis—University of Manchester, 2018.
6. Chang, R.J. Crude Oil to Chemicals—Industry Developments and Strategic Implications—Presented at Global Refining & Petrochemicals Congress (Houston, USA), 2018.
7. Lambert, N., Ogasawara, I., Abba, I., Redhwi, H., Santner, C. *HS-FCC for Propylene: Concept to Commercial Operation*, PTQ Magazine, 2014.
8. Mukherjee, M., Vadhri, V., Revellon, L. *Step-Out Propane Dehydrogenation Technology for the ^{2}1st Century*, The Catalyst Review, 2021.
9. Oyekan, S.O. *Catalytic Naphtha Reforming Process*. 1st edition, CRC Press, 2019.
10. Silva, M.W. *More Petrochemicals with Less Capital Spending*, PTQ Magazine, 2020.
11. Tallman, M.J., Eng, C., Sun, C., Park, D.S. *Naphtha Cracking for Light Olefins Production*, PTQ Magazine, 2010.
12. Vu, T., Ritchie, J. *Naphtha Complex Optimization for Petrochemical Production*, UOP Company, 2019.
13. Youssef, F., Adrian, M.H., Wenzel, S. *Advanced Propane Dehydrogenation*, PTQ Magazine, 2008.
14. Zhou, T., Baars, F. *Catalytic Reforming Options and Practices*, PTQ Magazine, 2010.

5 Hydroprocessing Technologies

Over the years, in the face of rising pollution levels associated with technological development and the rise in petroleum consumption, environmental legislation has become increasingly severe.

Restrictions on SOx and NOx emissions induced the necessity for higher technology development that can allow reducing the contaminant levels in the petroleum derivates, mainly sulfur and nitrogen. Normally, the concentration of contaminants increases with the density of the petroleum derivate.

A lot of technologies were applied to reduce the contaminant levels in the petroleum derivates, for example, the kerosene treating with Clay, the adsorption of sulfur compounds over black carbon, and the recognized treatments Bender and Merox. The mentioned technologies show limitations, mainly when the concentration of contaminants is high.

The hydrotreating technology (treatment with hydrogen) has been studied by many researchers in the refining industry and academic sector over the decades, and currently, it is practically impossible to attend to the petroleum derivate specifications without these streams passing through the hydrotreating unit.

The hydrotreating process involves a series of chemical reactions between hydrogen and organic compounds containing the contaminants (N, S, O, etc.). According to the target contaminant of the hydrotreating, the process can be called hydrodesulfurization (removing S), hydrodenitrogenation (removing N), hydrodeoxygenation (removing O), or hydrodearomatization when the main objective is to saturate aromatic compounds, among others.

The most common hydrotreating forms are hydrodesulfurization (where the objective is to remove compounds like benzothiophene, dibenzothiophene, etc.) and hydrodenitrogenation (removing porphyrins, quinolines, etc.) These compounds, besides provoking emissions of SOx and NOx when burned, produce in the derivates acidity, color, and chemical instability.

The main chemical reactions associated with the hydrotreating process can be represented like so:

$$R\text{-}CH=CH_2 + H_2 \rightarrow R\text{-}CH_2\text{-}CH_3 \text{ (Olefin Saturation)}$$
$$R\text{-}SH + H_2 \rightarrow R\text{-}H + H_2S \text{ (Hydrodesulfurization)}$$
$$R\text{-}NH_2 + H_2 \rightarrow R\text{-}H + NH_3 \text{ (Hydrodenitrogenation)}$$
$$R\text{-}OH + H_2 \rightarrow R\text{-}H + H_2O \text{ (Hydrodeoxigenation)}$$

Where R represents a hydrocarbon

The hydrotreating process is normally conducted in fixed-bed reactors, and the most applied catalysts are cobalt (Co), nickel (Ni), molybdenum (Mo), and tungsten (W),

DOI: 10.1201/9781003291824-5

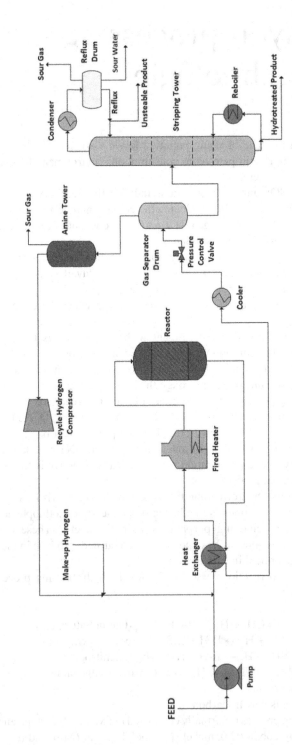

FIGURE 5.1 Basic Process Flow Diagram for Low-Severity Hydrotreating Processing Units

commonly in association with them and supported in alumina (Al_2O_3). The association Co/Mo is applied in reactions that need lower reactional severity, like hydrodesulfurization, while the catalyst Ni/Mo is normally applied in reactions that need higher severity, like hydrodenitrogenation and aromatic saturation.

Hydrotreating is applied in the finishing of the final products like gasoline, diesel, or kerosene or the intermediate step in the refining scheme in refineries to prepare feed charges to other processes like RFCC or hydrocracking (HCC), where the main objective is to protect the catalyst applied in these processes.

The basic process flow is like the various hydrotreating processes (hydrodesulfurization, hydrodenitrogenation, etc.). However, the process severity, determined by variables like hydrogen partial pressure, temperature, and catalyst, varies, and the contaminant removal is affected.

The hydrotreatment processing units are optimized, aiming for an equilibrium between cited operational variables because chemical reactions are exothermic, and the decontrolled raising in the temperature can affect the reactional equilibrium negatively. Besides, the sintering of the catalysts is possible. To minimize this risk, normally the hydrotreating reactors have points between the catalyst beds where hydrogen in lower temperature (quench lines) is injected to permit better control of the reactor temperature.

Figure 5.1 shows a typical arrangement for a hydrotreating processing unit with a single separating vessel.

The configuration with a single separating vessel is normally applied in lower-severity units, like hydrodesulfurization units. This arrangement is possible in this case because under reduced pressures, the difference between water and hydrocarbon properties is large, and the separation process needs to reduce contact areas, so a single vessel can realize the separation process.

Higher-severity units, like processing units dedicated to treating unstable streams (light cycle oil, coke gas oil, etc.) or with the objective to remove nitrogen or aromatic saturation, operate with two separating vessels, as presented in Figure 5.2.

In this case, the difference between water and hydrocarbon properties is small, and the phase separation process needs a higher interface area, so two separating vessels are applied—one under high pressure, where the separation among liquid and gaseous phases (H_2, H_2S, NH_3 and light hydrocarbons) occurs, and other under low pressure, where the separation between aqueous and hydrocarbon phases is promoted, apart from the separation of the remaining gases.

For lower-severity units, the temperatures applied are about 300–350°C and pressures vary between 20 and 40 bar, in addition to lower residence times. Units with high severity operate under temperatures of 350–400°C, and pressures vary from 40 to 130 bar.

Like aforementioned, great effort was employed in hydrotreating technology development. However, technology licensors like Axens, UOP, ExxonMobil, Lummus, Haldor Topsoe, and Albemarle, among others, still invest in research to improve the technology, mainly in the development of new arrangements that can minimize the hydrogen consumption (high-cost raw material) and that apply lower-cost catalysts that are more resistant to the deactivation process.

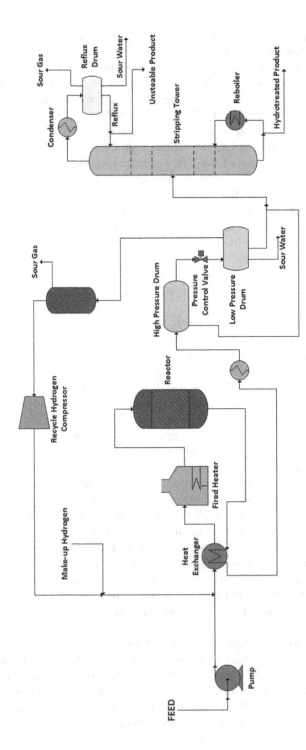

FIGURE 5.2 Basic Process Flow Diagram for High-Severity Hydrotreating Processing Units

5.1 NAPHTHA HYDROTREATING TECHNOLOGIES

5.1.1 Coker Naphtha Hydrotreating

Delayed coking is one of the most employed technologies dedicated to raising the recovery of high-added-value streams from crude oil. In this process, the vacuum residue from the vacuum distillation tower is converted into streams that can be used to produce middle and light derivatives through a controlled thermal cracking route.

One of the value streams produced by delayed coking process is the coker naphtha. This stream presents a boiling range close to 40–220°C. Due to being produced through a thermal cracking process, coker naphtha has high contaminant content (close to 2% sulfur and 500 ppm of nitrogen) and high olefin concentration (which can reach 50% in volume), which makes these streams chemically unstable (tendency to form gum).

These characteristics make the previous treatment of coker naphtha necessary before the destination to the gasoline or diesel (heavy naphtha) refinery pool. In this case, the appropriate treatment is a hydrotreating step.

Coker naphtha hydrotreating is a great challenge to the refining industry. The presence of a high quantity of olefins and contaminants leads to fouling and consequently raises the pressure drop in the catalytic beds. Furthermore, the presence of silicon-based compounds in the coker naphtha provokes the deactivation of the hydrotreating catalysts. The silicon-based compounds are added to the delayed coking units as anti-foam agents. The high olefin concentration, mainly diolefins, makes temperature control very hard in the hydrotreating reactors.

Due to these difficulties, the coker naphtha hydrotreating units are designed with three main objectives: diolefin saturation, removal of silicon-based compounds, and finally, reduction of contaminant content (hydrodesulfurization and hydrodenitrogenation). Figure 5.3 shows a process flow scheme for a typical coker naphtha hydrotreating unit.

The feed stream is fed to a hydrotreating rector aim of promoting diolefin saturation. This reactor operates under relatively low temperatures (150–240°C). Due to the high heat production during diolefin saturation, in some cases the refiners can blend coker naphtha with straight-run naphtha to reduce the heat production in this step. It's a good strategy to control the reactor temperature. The effluent of this reactor is fed to a second reactor that aims to remove silicon contaminants. The catalysts applied in this step have a greater porosity to avoid quick pressure drop elevation and plugging the reactor.

Next, the effluent of the second reactor is fed to another hydrotreating reactor, where the hydrodesulfurization and hydrodenitrogenation reactions take place. The product from this reactor is sent to a separation system and directed to consumers according to the refining scheme adopted by the refiner. Normally, the hydrotreated coker naphtha is feed-streamed to the catalytic reforming unit, but in some cases, it can be directed directly to the gasoline pool or commercialized as a petrochemical intermediate.

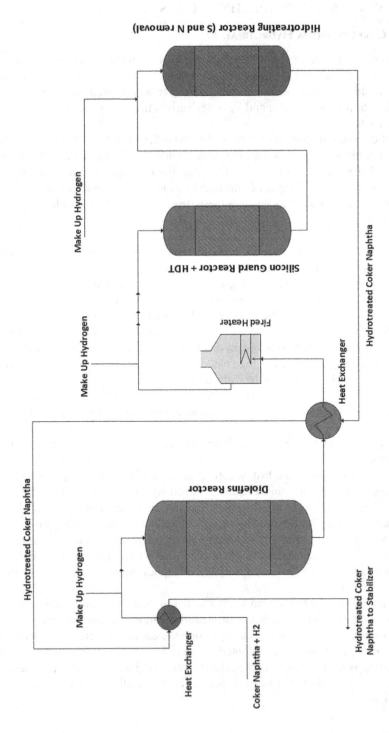

FIGURE 5.3 Typical Process Arrangement for Coker Naphtha Hydrotreating Unit

In refineries that don't have coker naphtha hydrotreaters in their refining scheme, it's common to feed the coker naphtha to FCC unit or for high-severity hydrotreating units (heavy coker naphtha) to compose the diesel pool of the refinery.

Due to the strategic character of this technology, some refining technology licensors such as Axens, UOP, Haldor Topsoe, and Shell Company developed and commercialized process technologies dedicated to treating coker naphtha. As cited earlier, the catalysts are fundamental to the success of the coker naphtha hydrotreating process. The catalyst needs present high activity and be resistant to the high contaminant content (sulfur, nitrogen, and silicon). Some companies have dedicated their efforts to developing catalytic systems capable of attending to these requirements. Examples of these technologies are the Start system (by W.R. Grace), the Unity system (by UOP), and the Sentry catalysts (by Criterion Catalysts).

Coker naphtha hydrotreating is capable of raising the refiner profitability once it converts a residual stream with a difficult destination into derivatives with high added value (such as catalytic reforming feed stream) and petrochemical naphtha. Faced with the current scenario of the increased installation of delayed coking units in the refining industry, the coker naphtha hydrotreating technology acquires a fundamental role in guaranteeing competitiveness to the refiners.

5.1.2 FCC Naphtha Hydrotreating Technologies

The technology challenge is especially hard to treat the naphtha from the FCC unit dedicated to producing gasoline, one of the most consumed crude derivatives in the world market.

The use of crude oil, increasingly heavier and consequently with higher contaminant content, mainly sulfur, further increased the pressure of the regulatory agents on the refiners in the sense that they reduce the contaminant content in the derivatives, especially diesel and gasoline. The main technology used to reduce the contaminant content in crude oil derivatives is hydrotreating.

The conventional hydroprocessing scheme is common and widely applied in the refining industry. However, to produce high-quality gasoline, its use is limited as hydrotreating reactions fatally lead to olefin saturation, which is responsible for the high-octane number in the final gasoline. Straight-run naphtha is normally directed to conventional hydrotreating because it has a reduced olefin content at the same time, which presents reduced sulfur content. In this case, the mild hydrotreating process is effective. On the other hand, naphtha from FCC tends to show high sulfur concentration and high olefin content. The processing of this stream in conventional hydrotreating units would be poorly effective.

The main chemical reactions associated with cracked naphtha hydrotreating can be represented as follow:

$$R\text{-}CH=CH_2 + H_2 \rightarrow R\text{-}CH_2\text{-}CH_3 \text{ (Olefin Saturation) (1)}$$
$$R\text{-}SH + H_2 \rightarrow R\text{-}H + H_2S \text{ (Hydrodesulfurization) (2)}$$

Where R is a hydrocarbon

As aforementioned, in the case of naphtha hydrotreating gasoline, the objective is to minimize the reaction (1) and maximize the yield of the reaction (2) to keep the high-octane number while the sulfur content is reduced, making the naphtha more environmentally friendly.

Over the last decades, technology process licensors devoted their efforts to developing technologies capable of achieving these objectives. UOP commercializes the SelectFining technology, which applies fixed-bed reactors to promote selective hydrotreating of naphtha, leading to the unstable compound (diolefin) saturation in the first reactor and the heavier olefin (higher sulfur content) saturation in the second reactor through the use of an adequate catalyst.

Nowadays, one of the most used technologies to reduce the sulfur content in the cracked naphtha is the Prime-G+ process, developed by Axens. The process takes advantage of the tendency of the sulfur to concentrate in the heavier fractions of the cracked naphtha. Therefore, it is carried out a feed stream fractionation before the hydrotreating step, as presented in Figure 5.4, where FRCN = full-range cracked naphtha, HCN = heavy cracked naphtha, LCN = light cracked naphtha, and SHU = selective hydrotreating unit (diolefin reactor).

The feed stream is fed to a hydrotreating reactor aim of promoting diolefin saturation, and then the stream is separated into light and heavy fractions in a distillation tower. While the light naphtha is recovered in the top, the heavy fraction is removed from the bottom of the column and sent to a selective hydrotreating section, leading to a minimum octane loss. In the sequence, the hydrotreated naphtha is separated in a stabilizer column to remove light compounds, the bottom product is mixed with the light fraction, and the final product is directed to the refinery gasoline pool.

Another technology that applies selective hydrotreating to reduce the sulfur content in the cracked naphtha is the Scanfinning process developed by ExxonMobil. In this case, fixed-bed reactors are applied without feed stream fractionation. Regarding the catalysts dedicated to selective hydrogenation of FCC naphtha, we can highlight the product HyOctane, developed by Haldor Topsoe.

The GT-DeSulph process, developed by GTC Engineering, applies the extractive distillation principle associated with selective hydrotreating. The use of aromatic solvents allows the removal of thiophenic compounds from cracked naphtha.

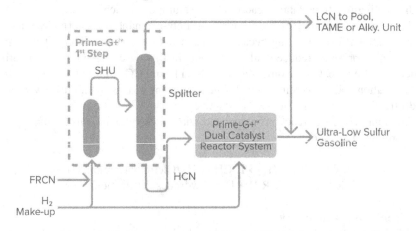

FIGURE 5.4 Basic Process Flow Diagram for the Prime-G+ Process by Axens (with Permission)

Hydroprocessing Technologies

Another interesting route adopted by the licensors is the non-selective hydrotreating followed by an octane index recovery step. The ISAL technology, by UOP, employs this principle. In this case, through an adequate catalyst, the olefin saturation occurs, and the molecular rearrangement is posteriorly carried out to recover the octane index lost in the first step.

The S-Zorb process, developed by ConocoPhillips, applies the concept of chemical adsorption of the sulfur compounds on zinc oxide (ZnO) and has a continuous regeneration system of the adsorbent through the controlled burn of the adsorbed sulfur compounds.

FCC naphtha is a fundamental stream for the refiners to meet the automotive gasoline requirements (minimum octane number), and the adequate treatment of this process stream is necessary to ensure the adherence to environmental regulations at the same time in raising the profitability of the refiner once the cracked naphtha yield in FCC units can reach over than 50%.

5.1.2.1 A Special Challenge: Diene (Diolefin) Control

A key parameter in the cracked naphtha is the conjugated diene content, these compounds are chemically unstable and tend to undergo oxidation reactions leading to gum formation in the reactors in hydrodesulfurization units, leading to high pressure drop and lower operation campaigns.

The diene formation in FCC units occurs mainly in the transfer line between the reaction section and the fractionator. This phenomenon is caused by the absence of catalyst in this section, and due to the high temperature, thermal cracking occurs, favoring the diene production.

Due to these characteristics, the refiners normally control the diene production through the quench injection at some point post riser, aiming to reduce the temperature, avoiding the thermal cracking and diene formation. The literature quotes that it is possible to reduce close to 50% of the diene formation by applying a post-riser quench with light cycle oil (LCO). The main process licensors recommend the quench injection in the separator vessel (diluted phase) as the injection in the transfer line can lead to hydrocarbon condensation and coke deposition in the bottom of the main fractionator. The use of quench in the riser tends to raise the production of fuel gas and coke, which can limit FCC units in the machines like blowers and cold area compressors.

The diene concentration in the cracked naphtha is normally controlled through the maleic anhydride value (MAV) that is defined by a quantity of maleic anhydride (in mg) necessary to react with one gram of cracked naphtha through the Diels-Alder mechanism as presented in Figure 5.5.

Some refiners usually control the MAV below 4 mg/g, but it's important to consider that the diene concentration is not the unique parameter to keep a reliable operation in the cracked naphtha hydrodesulfurization unit. Control of the concentration of light compounds in the feed (C_4) is important to avoid sudden vaporization in the catalyst bed, leading to lower reaction temperatures in the selective hydrogenation unit that cause lower conversion, consequently raising the diolefin (diene) concentration in the main reaction section that can lead to gum formation, reducing the operation life cycle. The light concentration in the cracked naphtha is controlled through

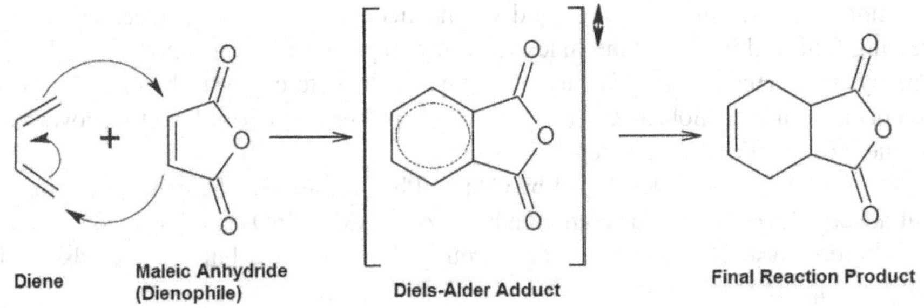

FIGURE 5.5 Diels-Alder Reaction for a Conjugated Diene

Source: Based on Solomons et al., 2018

the Reid vapor pressure. Another side effect of the diolefin concentration in FCC naphtha hydrotreaters is related to the polymerization in the catalyst bed, leading to higher pressure drop and shorter operating life cycle.

Faced with the current specifications of gasoline, the cracked naphtha hydrodesulfurization unit became fundamental to allow refiners relying on FCC units in their refining scheme to produce marketable gasoline. An unplanned shutdown of this unit can quickly lead to an FCC shutdown and, in extreme cases, the interruption of refinery operations.

In some cases, premature activity loss can be observed in the selective hydrotreating unit (SHU). This scenario can be overcome through lower temperature in the FCC unit (TRX), which can be compensated through a high activity of the FCC catalyst.

5.2 DIESEL HYDROTREATING UNITS

To comply with the new regulations, diesel production requires higher-severity units. Normally the straight-run diesel is processed with unstable streams (light cycle oil, coke gas oil, etc.) in high-severity units as presented in Figure 5.2, where it is possible to remove nitrogen or aromatic saturation. The unit operates with two separate vessels.

The high severity is required as, in the special case of middle distillates like diesel, contaminants like sulfides, thiophenes, and aromatics sulfur compounds like dibenzothiophenes are among the most difficult compounds to remove through hydrotreating. The hydrodesulfurization reactions are favored by higher temperatures, while the hydrodearomatization reactions are favored by higher hydrogen partial pressures. For this reason, the performance of diesel hydrotreating units requires a good balance between these variables, especially for units processing streams from FCC units, like light cycle oil (LCO), which presents high aromaticity and high sulfur content.

For higher-severity units like diesel hydrotreating, the difference between water and hydrocarbon properties is small, and the phase separation process needs a higher interface area, so two separating vessels are applied—one under high pressure, where the separation among liquid and gaseous phases (H_2, H_2S, NH_3, and light hydrocarbons) occurs, and other under low pressure, where the separation between aqueous and hydrocarbon phases is promoted, apart from the separation of the remaining gases. Units with high severity operate under the temperature of 330–380°C and pressures varying from 40 to 120 bar.

Due to the stricter limit of sulfur content in the diesel (<10 ppm), Ni/Mo is applied as catalysts to diesel hydrotreating units once this catalyst presents a higher activity to desulfurization reactions.

The principal process variables considered in diesel hydrotreating units are the total pressure, hydrogen partial pressure, makeup hydrogen purity, recycle gas rate, reactor temperature, and space velocity (liquid hourly velocity, LHSV). The space velocity defines the time required to achieve the desired performance of the reactions. This parameter can be defined as presented in Equation 1.

$$\text{LHSV (h}^{-1}) = \text{Feed Rate (m}^3\text{/h)} / \text{Catalyst Volume (m}^3) \tag{5.1}$$

The LHSV is a key parameter to hydrotreating units, not only to the design but also to the optimization of the unit, as it's possible to estimate the start of run (SOR) temperature and control the catalyst life cycle based on the end of run (EOR) temperature that is normally limited by the mechanical resistance of the material applied to reactors design. Typical diesel hydrotreating units present LSHV between 0,75 and 2,5 h^{-1}. In some designs, especially for deep hydrodesulfurization units, the reaction section is separated into two stages with the removal of H_2S and NH_3 between the reaction stages, aiming to minimize the deactivation effect of these gases over the catalyst. This can be especially attractive to hydrotreating units focused on producing ultra-low-sulfur diesel (ULSD).

As aforementioned, diesel is the crude oil derivative that had the most increased demand in the last decades. This derivative is mainly applied as a transportation fuel by vehicles equipped with diesel cycle engines. It is composed of hydrocarbons between C_{10} and C_{25}, with a boiling range of 150–380°C. Diesel ignition quality is measured through the cetane number that corresponds to a volumetric percentage of cetane (n-hexadecane) in a mixture with heptamethylnonane and burns with the same ignition quality of the analyzed diesel. The linear paraffinic hydrocarbons are the compounds that most contribute to the diesel ignition quality, raising the cetane number, while the presence of aromatics reduces this parameter and harms the ignition quality. Currently, the minimum cetane number of commercial diesel is 48.

Another important parameter controlled in the diesel is the plugging point that aims to control the content of linear paraffin, which tends to crystallize under low temperatures and harms the fuel supply to the engine. The plugging point is determined according to the weather conditions in the region of application. In Brazil, the plugging point is controlled in the range of 0–10°C. In colder regions, the cold flow properties tend to be a significant concern to refiners, especially those processing light and paraffinic crudes like north American shale oils. In these cases, the refiners normally install dewaxing beds in the hydrotreating reactors containing catalysts based on zeolites to promote the cracking of longer paraffin.

The diesel emissions control is carried out by managing the fuel density to control the content of heavy compounds, especially polyaromatics. Currently, the density of commercial diesel is controlled in the range of 830 to 865 kg/m^3. For ultra-low-sulfur diesel (ULSD), this parameter is controlled to be below 850 kg/m^3. In the last decades, there have been great efforts to reduce the environmental damage produced by diesel burn. Nowadays, environmental regulations require the commercialization of low-sulfur diesel with a maximum sulfur content of 10 ppm. However, in some markets, mainly in developing countries, commercialized diesel still has higher sulfur content (500 ppm), but this will change soon.

Great efforts were employed in the hydrotreating technology development. However, technology licensors like Axens, UOP, ExxonMobil, Lummus, Haldor Topsoe, and Albemarle, among others, still invest in research to improve the technology, mainly in the development of new arrangements that can minimize the hydrogen consumption (high-cost raw material) and that apply lower-cost catalysts that are more resistant to the deactivation process.

5.3 BOTTOM BARREL HYDROTREATING TECHNOLOGIES (RESIDUE UPGRADING)

The hard scenario faced by the whole oil and gas industry requires alternative actions to ensure maximum added value to crude oil. The downstream sector faces lower demand for crude oil derivatives and is put under pressure by the refining margins and competitiveness, mainly from refiners relying on refining hardware with low complexity and poor capacity of bottom barrel conversion and capable of producing lower yields of added-value derivatives. In the current scenario, the players of the downstream sector can consider processing heavier crude slates to improve their refining margins as these crudes present low cost when compared with lighter crude oil, or in some cases, there is a great availability of heavy crudes.

In this scenario, processing units capable of adding value to bottom barrel streams, improving the quality of crude oil residue streams (vacuum residue, gas oils, etc.), or converting them to higher-added-value products gain strategic importance, mainly in countries that have large heavy crude oil reserves, like Venezuela, Canada, and Mexico. These processing units are fundamental to comply with the environmental and quality regulations, as well as to ensure the profitability and competitiveness of refiners by raising the refining margin. The necessity to add value to bottom barrel streams increased even more after January 2020, when the IMO 2020 started, which requires a deep reduction in the sulfur content of the marine fuel oil (bunker) from 3,5% (m.m) to 0,5% (m.m).

Available technologies for processing bottom barrel streams involve processes that aim to raise the H/C relation in the molecule, either through reducing the carbon quantity (processes based on carbon rejection) or through hydrogen addition. Technologies that involve hydrogen addition encompass hydrotreating and hydrocracking processes, while technologies based on carbon rejection refer to thermal cracking processes like visbreaking, delayed coking and fluid coking, catalytic cracking processes like FCC, and physical separation processes like solvent deasphalting units.

Another fact that raises the relevance of the residue upgrading technologies is the growing demand for petrochemicals. This fact requires a higher bottom barrel conversion capacity to convert residual streams to petrochemical intermediates.

Hydroprocessing Technologies

The hydroprocessing of residual streams presents additional challenges when compared with the treating of lighter streams, mainly due to the higher contaminant content and residual carbon (RCR) related to the high concentration of resins and asphaltenes in the bottom barrel streams.

Higher metal and asphaltene content leads to a quick deactivation of the catalysts through a high coke deposition rate, catalytic matrix degradation by metals like nickel and vanadium, or even the plugging of catalyst pores produced by the adsorption of metals and high-molecular-weight molecules in the catalyst surface. For this reason, according to the content of asphaltenes and metals in the feed stream, more versatile technologies are adopted to ensure an adequate operational campaign and an effective treatment.

To demonstrate the mechanism of catalyst plugging, Figure 5.6 presents a scheme of reactant and product flows involved in a heterogeneous catalytic reaction as carried out in hydroprocessing treatments.

To carry out hydroprocessing reactions, the mass transfer of reactants to the catalyst pores, adsorption on the active sites to posterior chemical reactions, and desorption are necessary. In the case of bottom barrel stream processing, the high molecular weight and high contaminant content require a higher catalyst porosity to allow the access of these reactants to the active sites, allowing hydrodemetallization, hydrodesulfurization, hydrodenitrogenation, and so on. Furthermore, part of the feed stream can be in the liquid phase, creating additional difficulties to the mass transfer due to the lower diffusivity. To minimize the plugging effect, in fixed-bed reactors, the first beds are filled with higher-porosity solids without catalytic activity and act as filters to the solids present in the feed stream, protecting the most active catalyst from the deactivation (guard beds).

The process conditions are severer in the residue hydrotreating. The feed stream characteristics lead to a strong tendency of coke deposition on the catalyst, requiring

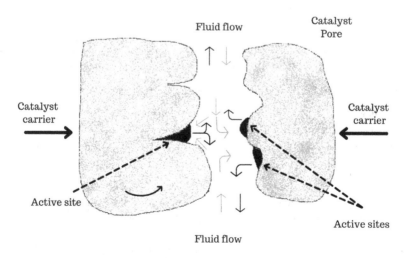

FIGURE 5.6 Reactant and Product Flows in a Generic Porous Catalyst

higher hydrogen partial pressures (until 160 bar to fixed-bed reactors) and higher temperatures (400–420°C).

Bottom barrel stream hydroprocessing can be applied to prepare the feed stream for other deep conversion processes like FCC and RFCC. It's also common to use high-severity hydrotreating processing units to reduce the contaminant content of the processing in hydrocracking units, with the objective of protecting the hydrocracking catalyst. Gas oil hydrotreating is very common in the preparation of feed stream to FCC units to control the content of sulfur, metals, and nitrogen, as well as promote the opening of aromatics rings that are refractory of the catalytic cracking reactions.

The hydrotreating process is normally conducted in fixed-bed reactors, and the most applied catalysts are cobalt (Co), nickel (Ni), molybdenum (Mo), and tungsten (W), commonly in association with them and supported in alumina (Al_2O_3). The association Co/Mo is applied in reactions that need lower reaction severity, like hydrodesulfurization, while the catalyst Ni/Mo is normally applied in reactions that need higher severity, like hydrodenitrogenation and aromatic saturation. For the hydrotreating of bottom barrel streams (vacuum gas oil, delayed coking gas oil, etc.), due to the higher severity needed, nickel-molybdenum (Ni-Mo) catalysts are applied.

Among the bottom barrel stream hydrotreating technologies, we can quote the Aroshift process (developed by Haldor Topsoe), the Unionfining process (developed by UOP), the Hyvahl technology (by Axens), and the RHU process (by Shell).

The residue hydroprocessing can also be realized through hydrocracking processing units according to the feed stream characteristics and the chosen refining configuration. Table 5.1 presents the main differences between the hydrotreating and hydrocracking processes.

Streams with higher contaminant content, especially metals, requires treatment by hydrocracking. As aforementioned, in some refining schemes, hydrotreating units can be applied to prepare the feed stream for hydrocracking units aiming to control the concentration of metals and nitrogen and protect the hydrocracking catalysts that normally have a high cost. Figure 5.7 presents a typical hydrocracking processing unit with two reaction stages.

TABLE 5.1
Hydrotreating and Hydrocracking Processes Comparison

Hydrotreating	Hydrocracking
Contaminant removal (S, N, O, metals, etc.) and C-C bond saturation	Contaminant removal (S, N, O, metals, etc.), cracking of C-C bonds, and reduction in molecular weight
Minimum cracking	High cracking rate
Low conversion (< 20%)	High conversion (>50%)
Feed stream preparation for conversion units—FCC/RFCC, catalytic reform, hydrocracking, etc.	Production of final products—transportation fuel (diesel and kerosene) and lubricants
Ni/Co/Mo typical catalysts	Ni/W/Pt/Pd typical catalysts (dual character)

Hydroprocessing Technologies

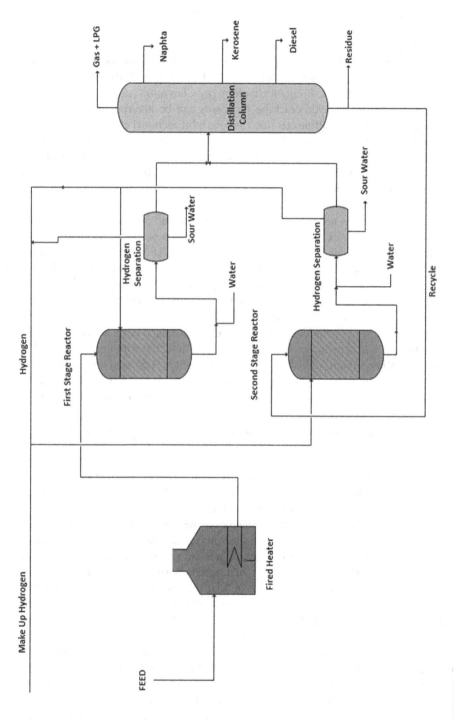

FIGURE 5.7 Typical Arrangement for Two-Stage Hydrocracking Units

The processing unit presented in Figure 5.7 relies on the intermediate separation of gases between the reaction stages. This configuration is adopted when the contaminant content (especially nitrogen) is high. In this case, the catalyst deactivation is minimized through the reduction of NH_3 and H_2S concentration in the reactors. Among the main hydrocracking process technologies available commercially, we can quote the H-Oil process (by Axens), the EST process (by Eni), the Uniflex process (by UOP), and the LC-Fining technology (by Chevron).

Catalysts applied in hydrocracking processes can be amorphous (alumina and silica-alumina) and crystalline (zeolites) and have bifunctional characteristics as the cracking reactions (in the acid sites) and hydrogenation (in the metal sites) occur simultaneously. The active metals used in this process are normally Ni, Co, Mo, and W, in combination with noble metals, like Pt and Pd. The hydrocracking process is normally conducted under severe reaction conditions with temperatures that vary from 300°C to 480°C and pressures between 35 and 260 bar.

A synergic effect is necessary between the catalyst and the hydrogen because the cracking reactions are exothermic and the hydrogenation reactions are endothermic, so the reaction is conducted under high partial hydrogen pressures, and the temperature is controlled to the minimum necessary to convert the feed stream. Despite these characteristics, the global hydrocracking process is exothermic, and the reaction temperature control is normally made through cold hydrogen injection between the catalytic beds and occurs in the hydrotreating processes.

Due to the severe operational conditions, the operational costs tend to be higher for the bottom barrel hydroprocessing units when compared with units dedicated to treating distillate streams (diesel, kerosene, and naphtha). The most intense hydrogenation process led to the most robust catalytic bed cooling systems (quench), higher hydrogen replacing rates, and complex phase separation systems (multiple stages).

5.4 ATMOSPHERIC RESIDUE DESULFURIZATION: A SPECIAL CASE

With the start of the validity of the new regulation regarding the quality parameters of marine fuel oil (bunker), some refiners and crude oil producers still question what will be the market behavior, faced with the new regulation. The IMO 2020 requires a deep reduction in the sulfur content of the marine fuel oil from the current 3,5% in mass to 0,5% in mass, leading to a necessity for changes in the production process of this derivative or higher control of sulfur content in the processed crude slate by the refiners.

To refiners with adequate bottom barrel processing capacity, the new regulation tends to be a great threat and can represent a good opportunity to raise the profitability, considering the competitive advantage that the high-complexity refining hardware gives to these refiners. The eventual devaluation of high-sulfur crude oil can experience due to the IMO 2020 can be translated into higher refining margins for refiners capable of processing these crudes.

One of the technologies that have been widely considered in the downstream industry in the IMO 2020 scenario is the desulfurization of atmospheric residue, aiming to allow not only the compliance with the new regulation but the quality improvement of the other derivatives and reliability of the downstream processing units like FCC or hydrocracking. As presented in Figure 5.8, the atmospheric residue corresponds to the bottom stream of the atmospheric crude oil distillation column.

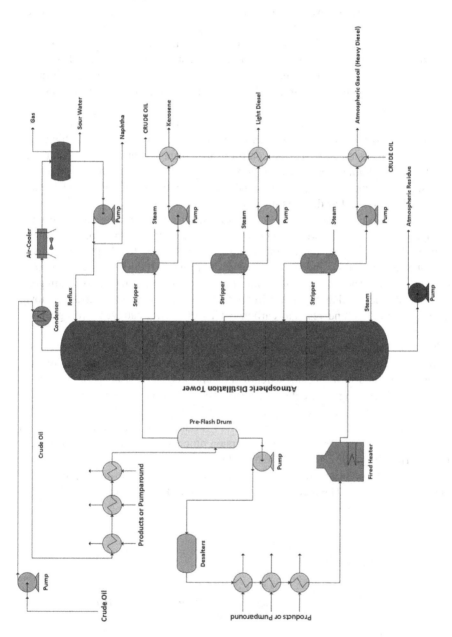

FIGURE 5.8 Typical Process Arrangement of Atmospheric Crude Oil Distillation Unit

Once heteroatoms like sulfur, nitrogen, and metals tend to concentrate in the heavier fractions of crude oil, the atmospheric residue drags a major part of the contaminants present in crude oil. Considering the current quality and environmental requirements over the derivatives, posterior treatments are required aiming to reduce the contaminant content (mainly sulfur and nitrogen) in the derivatives.

Before January 2020, the production of marine fuel oil (bunker) involves basically the dilution of vacuum residue (bottom barrel stream from vacuum distillation column) or deasphalted oil (to refiners that rely on a solvent deasphalting unit in the refining scheme) with lighter streams like LCO (light cycle oil) and gas oils, as presented in Figure 5.9.

The IMO 2020 makes necessary a better control of the sulfur content in the streams applied as diluents in the bunker production. For refiners with high bottom barrel conversion capacity, the control of the sulfur content in the vacuum residue through the atmospheric residue applying hydrodesulfurization minimizes the necessity of treatment of other streams and can prevent the use of the noblest streams like diesel and jet fuel as diluents in the bunker production.

The hydrodesulfurization process of atmospheric residue presents additional technological challenges when compared with the hydrotreating process applied to final derivatives like diesel and gasoline, considering the high contaminant content, mainly metals, and the residual carbon due to the high concentration of resins and asphaltenes in the feed stream. Beyond the sulfur removal, the main goal, the atmospheric residue hydrodesulfurization unit promotes the partial removal of metals, nitrogen, and residual carbon (CCR) through a catalytic hydrogenation mechanism.

The role of the atmospheric hydrodesulfurization unit in the refinery goes beyond allowing the production of low-sulfur fuel. In high-complexity refineries, the unit is applied as a feedstock treatment step to conversion units such as FCC/RFCC, hydrocracking, and delayed coking. The reduction of contaminant content and residual carbon promoted by the atmospheric residue hydrodesulfurization unit significantly raises the quality of derivatives produced by downstream units and raises the catalyst life cycle of deep conversion processes like FCC and hydrocracking, contributing to the reduction the operation costs.

The process conditions tend to be more severe in the case of atmospheric residue hydroprocessing. The feedstock characteristics lead to a strong tendency of coke

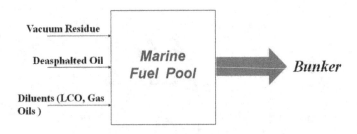

FIGURE 5.9 Marine Fuel Oil (Bunker) Production Process before IMO 2020

deposition over the catalyst requiring then higher hydrogen partial pressure (until 180 bar to fixed-bed reactors) and higher temperatures (380–420°C).

A typical atmospheric residue hydrodesulfurization unit can achieve 95% of conversion in hydrodesulfurization reactions and 98% in hydrodemetallization reactions. Furthermore, it's possible to achieve a reduction of 65% in residual carbon according to the employed technology. Normally, atmospheric hydrodesulfurization units rely on catalytic beds focused on removing metals, also called guard beds, aiming to protect the catalysts in the downstream reactors and improve the operational life cycle.

Due to the severe operating conditions, the operation costs of atmospheric residue desulfurization units are higher when compared with hydrotreating units dedicated to processing distillates (diesel, jet fuel, and naphtha). The most intense hydrogenation process leads to a necessity for more robust quenching systems of catalytic beds, higher hydrogen makeup rates, and more complex phase separation systems (multiple stages).

Despite the relatively high capital cost, the implementation of atmospheric residue hydrodesulfurization units allows refiners greater operational flexibility and high-sulfur oil producers and greater value addition to crude oil.

5.5 HYDROCRACKING TECHNOLOGIES

As presented in Table 5.1, for conversion rates above 50%, the hydroprocessing process can be called the hydrocracking process as it achieves a significant reduction in the molecular weight of the molecules during the process and not only the contaminant removal like in the less severe hydroprocessing units.

There are two ways to improve the quality of bottom barrel streams: the residue upgrading processes. The objective of a residue upgrading process is to raise the H/C ratio in the molecules, which can be achieved through carbon rejection or hydrogen addition.

Among the carbon rejection residue upgrading technologies, it's possible to quote the visbreaking, delayed coking, FCC, and solvent deasphalting technologies already presented in this book. The hydrogen addition route is represented by the bottom barrel hydroprocessing and hydrocracking technologies.

The hydrocracking process is normally conducted under severe reaction conditions with temperatures that vary from 300°C to 480°C and pressures between 35 and 260 bar. Due to process severity, hydrocracking units can process a large variety of feed streams, which can vary from gas oils to residues that can be converted into light and medium derivates with high added value.

Among the feed streams normally processed in hydrocracking units are the vacuum gas oil, light cycle oil (LCO), decanted oil, and coke gas oil. Some of these streams would be hard to process in FCC units because of the high contaminant content and the higher carbon residue, which quickly deactivates the catalyst. In the hydrocracking process, the presence of hydrogen minimizes these effects.

According to the catalyst applied in the process and the reaction conditions, hydrocracking can maximize the feed stream conversion in middle derivates (diesel and kerosene) and high-quality lubricant production (lower-severity process).

Catalysts applied in hydrocracking processes can be amorphous (alumina and silica-alumina) and crystalline (zeolites) and have bifunctional characteristics as the

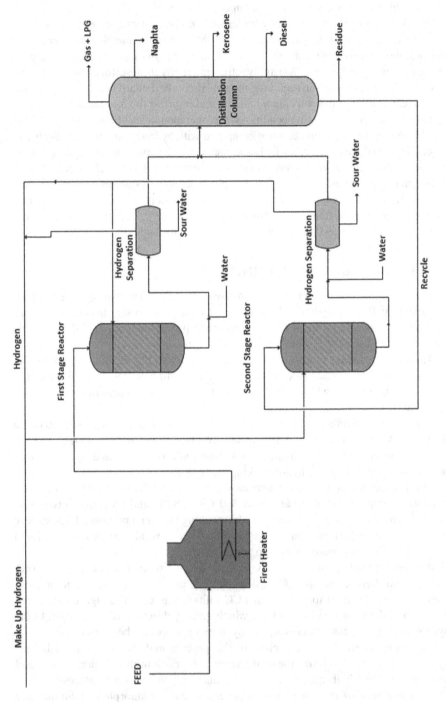

FIGURE 5.10 Basic Process Flow Diagram for Two-Stage Hydrocracking Units

Hydroprocessing Technologies

cracking reactions (in the acid sites) and hydrogenation (in the metal sites) occur simultaneously. The active metals used in this process are normally Ni, Co, Mo, and W, in combination with noble metals like Pt and Pd.

It's necessary for a synergic effect between the catalyst and the hydrogen because the cracking reactions are exothermic and the hydrogenation reactions are endothermic, so the reaction is conducted under high partial hydrogen pressures, and temperature is controlled at the minimum necessary to convert the feed stream. Despite these characteristics, the global hydrocracking process is exothermic, and the reaction temperature control is normally made through cold hydrogen injection between the catalytic beds.

Figure 5.10 shows a typical arrangement for a hydrocracking processing unit with two reaction stages dedicated to producing medium distilled products (diesel and kerosene).

According to the feed stream quality (contaminant content), hydrotreating reactors installed upstream of the hydrocracking reactors are necessary. These reactors act like guard beds to protect the hydrocracking catalyst.

The principal contaminant of hydrocracking catalysts is nitrogen, which can be present in two forms: ammonia and organic nitrogen.

Ammonia (NH_3), produced during the hydrotreating step, has a temporary effect of reducing the activity of the acid sites, mainly damaging the cracking reactions. In some cases, the increase of ammonia concentration in the catalytic bed is used as an operational variable to control the hydrocracking catalyst activity. Organic nitrogen has a permanent effect, blocking the catalytic sites and leading to coke deposits on the catalyst.

As in the hydrotreating cases (HDS, HDN, etc.), the most important operational variables are temperature, hydrogen partial pressure, space velocity, and hydrogen/feed ratio.

Depending on feed stream characteristics (mainly contaminant content) and the process objective (maximize middle distillates or lubricant production), hydrocracking units can assume different configurations.

For feed streams with low nitrogen content where the objective is to produce lubricants (partial conversion), it is possible to adopt a single-stage configuration without the intermediate gas separation. Produced during the hydrotreating step, this configuration is presented in Figure 5.11. The main disadvantage of this configuration is the reduction of the hydrocracking catalyst activity caused by the high concentration of ammonia in the reactor, but this configuration requires lower capital investment.

The application of the hydrocracking route to produce lubricating oils offers a great competitive advantage once the alternative routes, based on solvent extraction units, are capable of producing only Group I and II lubricating oils that present falling demand. Downstream markets without installed hydrocracking units can face difficulties in meeting the market demand for high-quality lubricants. An example is the Brazilian market.

The Brazilian domestic market of paraffinic oils is supplied by two refineries that apply the solvent route to produce lubricating oils and waxes for a variety of consumers, like the food and cosmetic industries, among others. The national lubricating

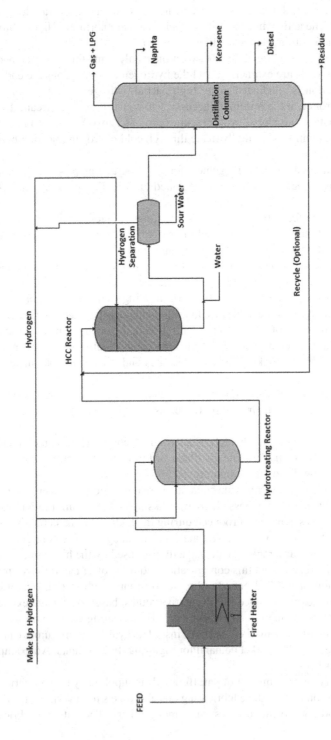

FIGURE 5.11 Typical Arrangement for Single-Stage Hydrocracking Units without Intermediate Gas Separation

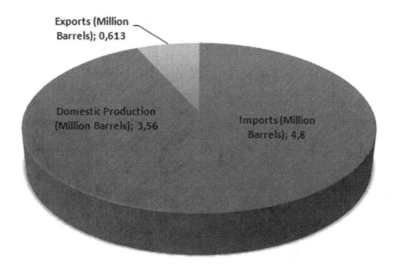

FIGURE 5.12 Balance of the Brazilian Market of Lubricating Oils in 2019

Source: ANP (2020)

production in 2019 was 3,5 million barrels. Additionally, the internal market is also supplied by some importers. According to data from the Brazilian Petroleum Agency (ANP), the internal consumption of lubricating oil reached 7,7 million barrels in 2019. Figure 5.12 shows the composition of the Brazilian market for lubricating oil in 2019.

Due to the accelerated technological development, especially in the automotive market, Group I lubricating oils tend to lose market in the next years. This fact tends to lead the refiners to look for capital investment to sustain their competitiveness in the lubricating oil market. Another side effect of lubricating oil producers based on solvent routes due to the loss of competitiveness is raising the imports to supply the internal market, leading to an external dependence on critical production input and negative effects on the balance of payments.

Normally for feed streams with low nitrogen content where the objective is to produce middle distillates (diesel and kerosene), the configuration with two reaction stages without intermediate gas separation is the most common. This configuration is shown in Figure 5.13.

As aforementioned, the disadvantage, in this case, is the high concentration of NH_3 and H_2S in the hydrocracking reactors, which reduces the catalyst activity.

The higher costly units are the plants with double stages and intermediate gas separation. These units are employed when the feed stream has high contaminant content (mainly nitrogen), and the refinery looks for the total conversion (to produce middle distillates). This configuration is presented in Figure 5.14.

In this case, the catalytic deactivation process is minimized by the reduction in the NH_3 and H_2S concentration in the hydrocracking reactor. It's important to consider the feedstock quality to define the better residue upgrading technology to the

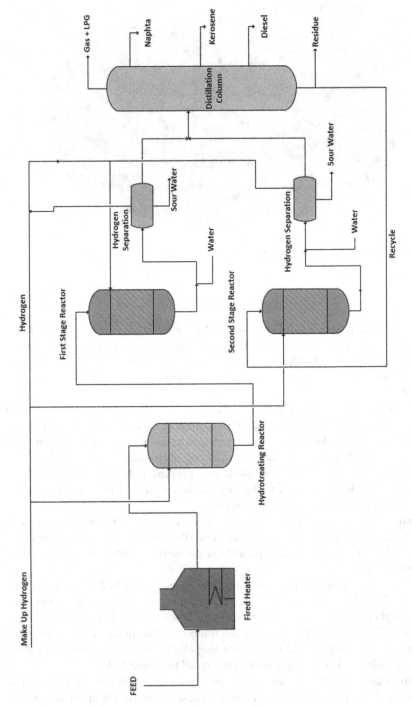

FIGURE 5.13 Typical Arrangement for Two-Stage Hydrocracking Units without Intermediate Gas Separation

Hydroprocessing Technologies

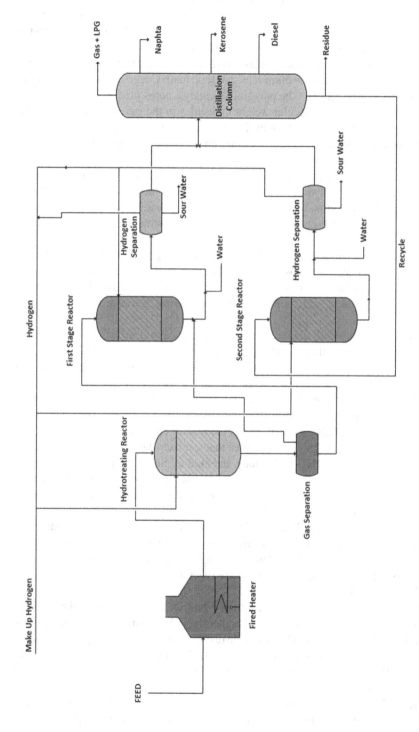

FIGURE 5.14 Typical Arrangement for Two-Stage Hydrocracking Units with Intermediate Gas Separation

refining hardware, as the hydroprocessing of residual streams presents additional challenges when compared with the treating of lighter streams, mainly due to the higher contaminant content and residual carbon (RCR) related to the high concentration of resins and asphaltenes in the bottom barrel streams.

Higher metal and asphaltene content leads to a quick deactivation of the catalysts through a high coke deposition rate, catalytic matrix degradation by metals like nickel and vanadium, or even the plugging of catalyst pores produced by the adsorption of metals and high-molecular-weight molecules in the catalyst surface. For this reason, according to the content of asphaltenes and metals in the feed stream, more versatile technologies are adopted to ensure an adequate operational campaign and an effective treatment.

As quoted earlier, hydrocracking units demand high capital investments, mainly to operate under high hydrogen partial pressures. It's necessary to install larger hydrogen production units, which is another costly process. However, faced with the increasing demand for high-quality derivates, the investment can be economically attractive.

The residue hydrocracking units have severity even greater than units dedicated to treating lighter feed streams (gas oils). These units aim to improve the residue quality either by reducing the contaminant content (mainly metals) like an upstream step to other processes, such as RFCC, or by producing derivates like fuel oil with low-sulfur content.

Residue hydrocracking demands an even greater capital investment than gas oil hydrocrackers because these units operate under more severe conditions, and furthermore, the operational costs are so higher, mainly due to the high hydrogen consumption and the frequent catalyst replacement.

Hydrocracking technologies have been widely studied over the years, mainly by countries with large heavy oil reserves, like Mexico and Venezuela. The main difference between the available technologies is the reactor characteristics.

Among the hydrocracking technologies that use fixed-bed reactors, the following can be highlighted: the RHU technology, licensed by Shell; the Hyvahl technology, developed by Axens; and the UnionFining process, developed by UOP. These processes normally operate with low conversion rates with temperatures higher than 400°C and pressures above 150 bar.

Technologies that use ebulliated bed reactors and continuum catalyst replacement allow higher campaign periods and higher conversion rates. Among these technologies, the most known are the H-Oil technology, developed by Axens, and the LC-Fining process by Chevron Lummus. These reactors operate at temperatures above 450°C and pressures until 250 bar.

An improvement in relation to ebulliated bed technologies is the slurry phase reactors, which can achieve conversions higher than 95%. In this case, the main available technologies are the HDH process (hydrocracking-distillation-hydrotreatment), developed by PDVSA-Intevep; the VCC process (Veba Combi-Cracking), developed by Veba Oil; and the EST process (Eni Slurry Technology), developed by the Italian state oil company Eni.

Despite the high capital investment and the high operational cost, hydrocracking technologies produce high-quality derivates and can make feasible the production of added-value products from residues, which is extremely attractive, mainly for countries that have difficult access to light oils with low contaminants.

In countries with a high dependency on middle distillates like Brazil (because of its dimensions and the high dependency on road transport), the high-quality middle distillate production from oils with high nitrogen content indicates that the hydrocracking technology can be a good way to reduce the external dependency of these products.

5.6 THE HYDROPROCESSING CATALYSTS

The hydrotreating catalysts are normally composed of metal sulfides of Group VI (W and Mo) and/or Group VIII (Ni and Co) carried by an oxide like alumina, zeolite, or silica-alumina. The most employed combinations in traditional hydrotreating processes are Co/Mo, Ni/Mo, and Ni/W. The combination Co/Mo is normally applied to hydrodesulfurization reactions once it presents less activity to harder reactions such as hydrodenitrogenation or aromatic saturation. In these cases, the catalyst selected is based on Ni/Mo combination, while the Ni/W catalyst is applied to deep hydroprocessing processes where the main objective is aromatic saturation. Normally, hydroprocessing reactors are filled with a combination of these catalysts aiming to optimize the performance and operating costs.

Some promoters can be added to the hydrotreating catalysts aiming to improve the performance in specific cases. Phosphorous is added to the Ni/Mo catalysts with the objective of improving the hydrodenitrogenation activity, and the fluorine is applied to improve the catalyst performance in cracking reactions through the higher acidity in the carrier. This is a great advantage in mild hydrocracking processes.

Catalysts applied in hydrocracking processes can be amorphous (alumina and silica-alumina) and crystalline (zeolites) and have bifunctional characteristics as the cracking reactions (in the acid sites) and hydrogenation (in the metal sites) occur simultaneously. The active metals used in this process are normally Ni, Co, Mo, and W, in combination with noble metals like Pt and Pd.

A synergic effect between the catalyst and the hydrogen is necessary because the cracking reactions are exothermic and the hydrogenation reactions are endothermic, so the reaction is conducted under high partial hydrogen pressures, and temperature is controlled to the minimum necessary to convert the feed stream. Despite these characteristics, the global hydrocracking process is exothermic, and the reaction temperature control is normally made through cold hydrogen injection between the catalytic beds.

For hydrocracking units, the catalyst activity is defined by the required temperature to reach the desired conversion, which is defined by Equation 5.2.

$$\text{Conversion (\%)} = [(1 - (\text{Fraction above TBP in the Product}) / (\text{Fraction above TBP in the Feed}))] \times 100 \qquad (5.2)$$

Where TBP is true boiling point, which represents the desired cut point defined by the refiner.

5.7 DEACTIVATION OF HYDROPROCESSING CATALYSTS

Hydroprocessing technologies allow the production of cleaner and better performance derivatives. At the same time, that makes possible the recovery of higher yields of added-value products from bottom barrel streams in the crude oil refining. For this reason, hydroprocessing technologies have become essential to the downstream industry in the last decades as it's practically impossible to produce marketable crude oil derivatives without at least one hydroprocessing step. To achieve the goal of ensuring maximum added value to the processed crude oil, refiners need an adequate hydroprocessing capacity in this refining hardware, especially those processing heavier crude oil, and one of the main concerns related to the operation of hydroprocessing units is the pressure drop in fixed-bed reactors.

Considering the relevance of hydroprocessing units to the refining hardware, unplanned shutdowns can lead to the impossibility of producing final derivatives like diesel, gasoline, and jet fuel, as well as the shutdown of other processing units according to the refining scheme. In some cases, the refiners are forced to reduce the throughput of the hydroprocessing unit to reduce the pressure drop in the reactors, leading to financial impacts. The growing trend of processing renewable raw materials in hydroprocessing units tends to increase; its relevance grows even more in the crude oil refineries.

In this sense, adequate management of pressure drop in hydroprocessing units is a key issue for refiners and requires great attention.

The main deactivation mechanisms of hydroprocessing catalysts are as follows:

- Metal deposition—related to feedstock characteristics and drag of contaminants
- Active phase sintering process—related to temperature and metal deposition
- Coking deposition—related to the processing conditions, feedstock characteristics, and operating issues; considered the only reversible deactivation process

The metal deposition is mainly affected by Ni, V, Pb, As, Si, Fe, and Na. Ni and Va can be present in heavier fractions of crude oil and plug the catalysts pore and act as coke precursors. Pb and As can react with the active phases (metal sulfides), leading to the sintering process and, consequently, the reduction of the active phase area. Pb is found in naphtha fractions, and the As can be found in all petroleum fractions.

Contamination by silicon normally occurs due to the injection of silicon-based compounds in the crude oil extraction step and in downstream processes like delayed coking units where anti-foaming agents are applied. The silicon acts by reducing the surface area and plugging the catalyst pore, leading to a severe activity reduction. The deactivation by Na is similar to the Si process. In hydrocracking processes, the feed contamination by Na is a great concern once the basic character of sodium promotes the neutralization of acid function of the hydrocracking catalysts, leading to a drastic reduction in the conversion (Equation 5.2).

Coking deposition is related to the condensation of high-weight molecules (heavier aromatics and asphaltenes) present in heavier feeds. The coke deposition is

Hydroprocessing Technologies

also related to dehydrogenation, cracking, and polymerization reactions of heavier fractions. The deactivation occurs through the plugging of catalyst pores, blocking the mass transfer from the hydrocarbon to the active phase.

The coking deposition also reduces the active surface area and is normally followed by metal deactivation, mainly to hydroprocessing units dedicated to treating bottom barrel streams.

The coking deposition process is positively affected by temperature and negatively affected by hydrogen partial pressure. For this reason, hydroprocessing units dedicated to processing heavier feeds operate under higher pressures with the main objective of protecting the catalysts that are responsible for a great part of the operating costs of the refiners.

The main causes of high pressure drop in hydroprocessing reactors are the internals like distributors and trays, particulates that are normally dragged with the feedstock, organic species like olefins and asphaltenes, and the coking deposition related to low hydrogen partial pressure, inadequate distribution, or hot points in the catalyst bed. Nowadays, the increasing usage of renewable raw materials in hydrotreating reactors calls for even more attention due to the higher heat release, the concentration of chemically unstable components, and higher total acid number.

Among the available strategies to mitigate the pressure drop issue in fixed-bed hydroprocessing reactors, it's possible to say the following:

- Filtration of the feedstock: This strategy is especially important to feed on delayed coking units due to the presence of coke particulates.
- Antifouling dosage in the hydroprocessing unit: The main objective here is to control the corrosive process, avoiding the drag of corrosion material to the reactors.
- Sacrificial catalyst: This strategy is applied mainly in hydrotreating units dedicated to processing bottom barrel streams. It uses a high-porosity catalyst to act as a filter, retaining particulates and contaminants on top of the catalyst bed.
- Grading catalyst: The grading is applied to retain the contaminants in the first section of the bed through the application of non-active material.

The size and shape of the catalyst particles have a great effect on the pressure drop in the hydroprocessing reactor, as well as the catalyst load strategy affects the pressure drop in the bed. To improve the characteristics of the catalyst, Criterion developed the ATX catalyst shape, which, among other characteristics, can minimize the pressure drop in the catalyst bed. In dense load, one of the critical parameters is to control the load speed to avoid the catalyst cracking during the load, raising the fine production.

During the startup of hydroprocessing units, it's important to analyze the procedures to avoid a great quantity of liquid in the catalyst beds during the startup. The high quantity of liquid can vaporize abruptly during the final steps of the startup, leading to the catalyst broken and high pressure drop.

As aforementioned, controlling the catalyst life cycle is a key issue to refiners, and one of the main strategies adopted in the last few years is the use of guard beds in hydroprocessing catalysts to protect the catalysts, ensuring a longer and most profitable operating campaign.

The main objective of the guard bed is to protect the main and active catalyst against the following:

- Particulates from the feedstock that can be dragged like sediments, catalysts powder, and corrosion products that are capable of producing physical fouling
- Heavier hydrocarbons capable of leading to coking deposition
- Chemical unstable hydrocarbons capable of producing gum, like olefins and diolefins
- Metals and catalyst poisons like Ni, V, Fe, Si, and Na

As aforementioned, due to the higher concentration of contaminants, the guard beds are most common in hydroprocessing units dedicated to processing heavier feedstocks, as quoted previously. Normally, a grading strategy is applied in the catalyst bed to establish a staggering pore diameter and activity to the catalysts, keeping the catalysts at the top more resistant to the contaminants acting as a filter, protecting the most active catalyst in the bottom section.

The guard bed will be responsible for controlling the contaminant content (mainly metals) to the next catalyst sections and reducing the carbon residue (CCR) and particulate concentration, keeping the activity and improving the life cycle of the hydroprocessing unit.

Among the most known catalyst protection technologies available in the market, we can mention the CatTrap technology, developed by Crystaphase; this technology uses a ceramic bed acting as a filter to particulate materials, controlling especially the pressure drop in the catalyst bed.

For units dedicated to treating bottom barrel streams, the hydroprocessing catalyst needs present high activity and be resistant to the high contaminant content (sulfur, nitrogen, and silicon). Some companies have dedicated their efforts to developing catalytic systems capable of attending to these requirements. Examples of these technologies are the Start system, by Advanced Refining Technologies (ART); the Unity system, developed by UOP; the Sentry catalysts by Criterion Catalysts; and the TK-449 Silicon Trap, by Haldor Topsoe. Figure 5.15 presents a comparative study developed by Haldor Topsoe related to the improvement of the cycle length of a naphtha hydrotreating unit, applying grading particles to control the contaminant content in the main catalyst.

The increasing relevance of hydroprocessing technologies to the downstream industry requires even more attention from refineries to keep operations profitable and reliable in these units. The guard bed technologies have an important role in allowing the achievement of this goal. As presented in Figure 5.15, these technologies can improve in a significant manner the operational life cycle of hydroprocessing units.

Hydroprocessing technologies have become essential to refiners in the last decades as it is practically impossible to produce marketable crude oil derivatives without at least one hydroprocessing step, even for refiners processing lighter crudes.

Hydroprocessing Technologies

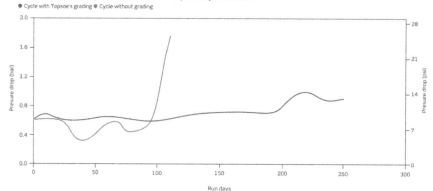

FIGURE 5.15 Cycle Length Improvement in a Naphtha Hydrotreating Unit with Catalyst Grading (Haldor Topsoe Company, with Permission)

Hydroprocessing units have a fundamental role in the downstream industry not only in the economic sustainability of the industry but also in keeping the environmental impact of crude oil derivatives under acceptable levels. In this sense, adequate management of hydroprocessing catalysts is a key factor in ensuring lower operating costs and competitiveness for refiners in the downstream market, and the guard bed technologies can offer an attractive route to ensure better performance and a larger operating life cycle for hydroprocessing units. As discussed previously, control of the pressure drop in hydrotreating reactors is one of the most important challenges to reach profitable and reliable operations. On the other side, there are optimization routes and technologies capable of ensuring adequate performance.

BIBLIOGRAPHY

1. Refining and Petrochemicals. *Encyclopedia of Hydrocarbons (ENI)*. Volume II, Refining and Petrochemicals, 2006.
2. Gary, J.H., Handwerk, G.E. *Petroleum Refining: Technology and Economics*. 4th edition, Marcel Dekker, 2001.
3. Leliveld, B., Toshima, H. *Hydrotreating Challenges and Opportunities with Tight Oil*, PTQ Magazine, 2015.
4. Robinson, P.R., Hsu, C.S. *Handbook of Petroleum Technology*. 1st edition, Springer, 2017.
5. Speight, J.G. *Heavy and Extra-Heavy Oil Upgrading Technologies*. 1st edition, Elsevier Press, 2013.
6. Zhu, F., Hoehn, R., Thakkar, V., Yuh, E. *Hydroprocessing for Clean Energy—Design, Operation, and Optimization*. 1st edition, Wiley Press, 2017.
7. Solomons, T.W.G., Fryhle, C.B., Snyder, S.A. *Organic Chemistry*. Volume 2. 12th edition, Rio de Janeiro, LTC Press, 2018.
8. Ancheyta, J., Speight, J.G. *Hydroprocessing of Heavy Oils and Residua*. 1st edition, CRC Press, 2007.
9. Seddon, D., Zhang, B. *Hydroprocessing Catalysts and Processes*. 1st edition, World Scientific Press, 2018.

6 Lubricating Production Refineries

A major part of the refining hardware in the world is focused on producing transportation fuels, but one of the most important crude oil derivatives is the lubricating base oil, which comprises around 85% of a final lubricant formulation.

Like others petroleum derivates, economic and technological developments have been required for the production of lubricating oils with higher quality and performance, moreover with lower contaminant content.

The main quality requirements for lubricating oils are viscosity, flash point, viscosity index (viscosity changes with temperature), fluidity point, chemical stability, and volatility.

According to American Petroleum Institute (API), the lubricating base oils can be classified as described in Table 6.1.

The lube oils from Groups II, III, and IV have higher quality than base oils from Group I. The content of contaminants like sulfur and unsaturated compounds are significantly reduced. Moreover, the viscosity index is superior for Groups II, III, and IV.

The first step in the lubricant production process is vacuum distillation of atmospheric residue obtained like the bottom product in the atmospheric distillation processes. For vacuum distillation units dedicated to producing lubricating fractions, the fractionating needs a better control than in the units dedicated to producing gas oils to fuel conversion. The objective is to avoid thermal degradation and to control the distillation curve of the side streams. A typical arrangement for vacuum distillation units to produce lubricating fractions is presented in Figure 6.1. A secondary vacuum distillation column is necessary when it is desired to separate the heavy neutral oil stream from vacuum residue.

TABLE 6.1
Lubricating Base Oils Classification according to API (American Petroleum Institute)

Group	Typical Production Process
I	Solvent extraction
II	Hydrocracking/hydrotreating or hydrocracking + solvent extraction
III	Hydrocracking/hydrotreating
IV	Synthetic

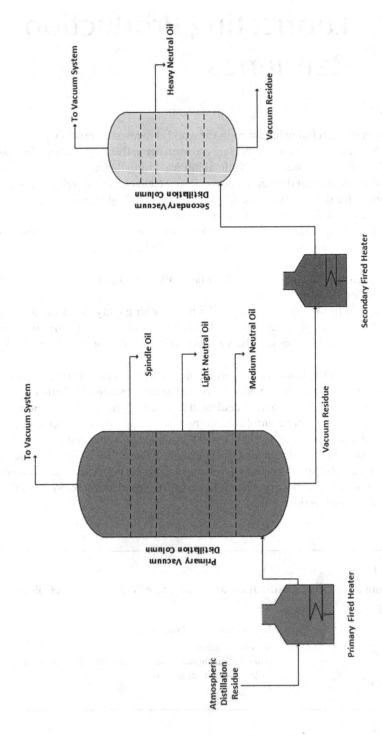

FIGURE 6.1 Typical Arrangement for Vacuum Distillation Process to Lubricating Oil Production

Lubricating Production Refineries

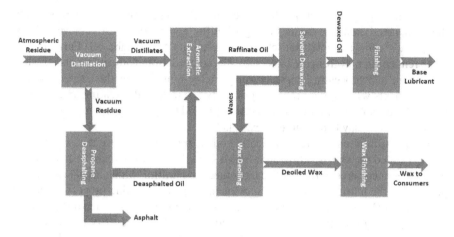

FIGURE 6.2 Processing Scheme for Base Lubricating Oil Production through Solvent Route

In lubricating production units based on the solvent route, the following steps are basically physical separation processes with the objective to remove from the process streams the components which can harm the desired properties of base oils, mainly the viscosity index and chemical stability.

Figure 6.2 shows a block diagram corresponding to the process steps to produce base lubricating oils through the solvent extraction route.

As aforementioned, in the vacuum distillation step, the fractionating quality obtained between the cuts is critical for these streams to reach the quality requirements like flash point and viscosity. After the vacuum distillation step, the side cuts are pumped to the aromatic extraction unit, and the vacuum residue is sent to the propane deasphalting unit. The propane deasphalting process seeks to remove from vacuum residue the heavier fractions, which can be applied as lubricating oil. The propane deasphalting units dedicated to producing lubricating oils apply pure propane-like solvent because this solvent has higher selectivity to remove resins and asphaltenes from deasphalted oil.

In the aromatic extraction step, the process streams are put in contact with solvents selective to remove aromatic compounds, mainly polyaromatics. The main objective in removing these compounds is the fact that they have a low viscosity index and low chemical stability. This is strongly undesired in lubricating oils. As the nitrogen and sulfur compounds are normally present in the polyaromatic structures, in this step a major part of the sulfur and nitrogen content of the process stream is removed. The solvents normally applied in the aromatic extraction process are phenol, furfural, and N-methyl pyrrolidone.

The subsequent step is to remove the linear paraffin with high molecular weight through solvent extraction. This step is important because these compounds harm the lubricating oil's flow at low temperatures. A typical solvent employed in the solvent dewaxing units is methyl-isobutyl-ketone (MIK), but some process plants apply toluene and/or methyl-ethylketone for this purpose.

After removing the paraffin, the lubricating oil is sent to the finishing process. In this step, heteroatom compounds (oxygen, sulfur, and nitrogen) are removed. These compounds can give color and chemical instability to the lube oil. Furthermore, some remaining polyaromatic molecules are removed. Some process plants with low investment and processing capacity apply a clay treatment in this step. However, modern plants with higher processing capacity use mild hydrotreating units. This is especially important when the petroleum processed has higher contaminant content. In this case, the clay bed saturates very quickly.

The paraffin removed from lubricating oils is treated to remove the excess oil in the unit called the wax deoiling unit. In this step, the process stream is submitted to reduced temperatures to remove the low branched paraffin, which has a low melting point. Like the lubricating oils, the subsequent step is a finishing process to remove heteroatoms (N, S, O) and saturate polyaromatic compounds. In the paraffin's case, generally, a hydrotreating process is applied with sufficient severity to saturate the aromatic compounds that can allow to reaching the food grade in the final product. As cited earlier, the solvent route can produce Group I lubricating oils. However, lube oils employed in severe work conditions (large temperature variation) need to have higher saturated compound content and higher viscosity index. In this case, it is necessary to use the hydrorefining route.

In the lubricating oil production by hydrorefining, the physical processes are substituted by catalytic processes, basically hydrotreating processes. Figure 6.3 shows a block diagram of the processing sequence to produce base lube oils through the hydrorefining route.

In this case, the fractionating in the vacuum distillation step has more flexibility than in the solvent route. As the streams crack in the hydrocracking unit, another distillation step is necessary.

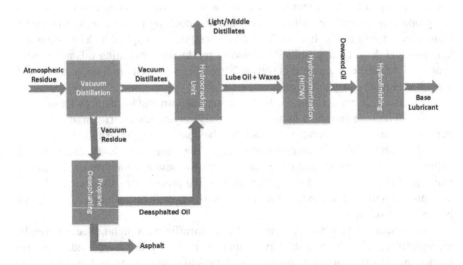

FIGURE 6.3 Processing Scheme for Base Lubricating Oil Production through Hydrorefining Route

After the vacuum distillation and propane deasphalting steps, the process streams are sent to a hydrotreating unit. This step seeks to saturate polyaromatic compounds and remove contaminants like sulfur and mainly nitrogen, which is a strong deactivation agent for the hydrocracking catalyst.

In the hydrocracking step, the feed stream is cracked under controlled conditions, and chemical reactions like dehydrocyclization and aromatic saturation occur, which give the process stream adequate characteristics for their application as lubricants. The following step, hydroisomerization, seeks to promote isomerization of linear paraffin (which can reduce the viscosity index), producing branched paraffin.

After the hydroisomerization, the process stream is pumped to hydrofinishing units to saturate the remaining polyaromatic compounds and remove heteroatoms. In the hydrofinishing step, the water content in the lube oil is controlled to avoid turbidity in the final product.

In comparing the lubricant production routes, it can be observed that the hydrorefining route gives more flexibility in relation to the petroleum to be processed. As the solvent route uses basically physical processes, it is necessary to select crude oils with higher paraffin content and low contaminant content (mainly nitrogen) for the processing, which can be a critical disadvantage in a geopolitically unstable scenario. The main disadvantage of the solvent route, when compared with the hydrorefining route, is that the solvent route can produce only Group I lubricating oils. This can limit its application to restricted consumer markets, which can reflect in the economic viability.

Another solvent route disadvantage is the solvents applied, which can cause environmental damage and needs special security requirements during the processing. The production of low-added-value streams like aromatic extract is another disadvantage.

Advantages of the solvent route over the hydrorefining route can be cited as lower capital investment and the fact that the solvent route produces paraffin, which can be directed to the consumer market as products with higher added value.

According to the recent forecasts from McKinsey & Company, the global market for lubricants is growing at an annual rate of around 4% and can reach a total value of $166 billion in 2025. The higher added value of lubricants in comparison with transportation fuel accompanied by the trend of reduction in transportation fuel demand indicates an attractive alternative to refiners with adequate refining hardware to improve their revenues and competitiveness in the downstream market.

As described previously, comparing the lubricant production routes can be observed that the hydrorefining route gives more flexibility in relation to the petroleum to be processed. As the solvent route uses basically physical processes, it is necessary to select crude oil with higher paraffin content and low contaminant content (mainly nitrogen) for the processing, which can be a critical disadvantage in a geopolitically unstable scenario. The main disadvantage of the solvent route, when compared with the hydrorefining route, is that the solvent route can produce only Group I lubricating oils. This can limit its application to restricted consumer markets, which can reflect in the economic viability. Some forecasts (Statista, 2020) indicate that the market for Group I lubricating oils is in contraction, and a significant reduction in the demand for Group I base oils is expected, leading to a great competitive loss to refiners relying on base oil production exclusively through solvent routes.

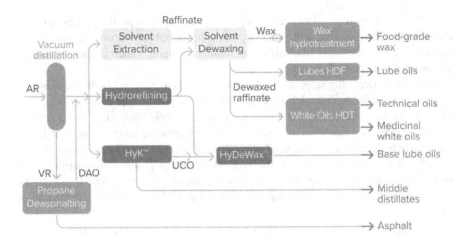

FIGURE 6.4 Block Diagram for the Hydewax Technology from Axens (with Permission)

Another solvent route disadvantage is the solvents' application, which can cause environmental damage and needs special security requirements during the processing. The production of low-added-value streams like aromatic extract is another disadvantage.

Advantages of the solvent route over the hydrorefining route can be cited as lower capital investment and the fact that the solvent route produces paraffin, which can be directed to the consumer market as products with higher added value. Figure 6.4 presents the block diagram for the Hydewax technology developed by Axens, which can help the refiners to maximize the added value through base lube oils and paraffin production. In this case, a synergy between solvent and hydrorefining production routes is applied.

6.1 CLOSING THE SUSTAINABILITY CYCLE: USED LUBRICATING OIL RECYCLING

Currently, one of the main challenges is to minimize natural resource consumption. In this sense, recycling processes are fundamental for the sustainability of our current way of life.

Recycling or re-refining used lubricating oil meets a double role: eliminate a hazardous residue and reduce the necessity for the extraction of higher quantities of petroleum to produce base lubricating oils. Unlike other petroleum derivates (diesel, gasoline, and kerosene), lubricating oils are not consumed during their application. However, during the life cycle, lubricating oils are contaminated and undergo severe degradation, leading to the loss of chemical and physical properties.

Like other chemical processes, re-refining technologies were the target of intensive research aiming at process improvement in the sense of cost reduction and mainly environmental impact. Considering that the used lubricating oil is a

Lubricating Production Refineries

dangerous residue that needs adequate destiny, the technologies dedicated to recycling these residues can be faced as a very important part of the crude oil processing chain, mainly in the current scenario where the circular economy is increasingly relevant to society. In some markets, recycled lubricating oil has an important role in the lubricating market.

6.1.1 Used Lubricating Recycling Technologies

The first industrial process developed to recover the used lubricating oil is called the acid-clay process, or the Meiken process. A basic process flow diagram for the Meiken process is presented in Figure 6.5.

The first step in the process is passing the used oil on grids to remove coarse solids. Next, the feed stream is sent for a decanter to promote the separation of free water present in the oil.

The next step is the dehydration of the feed to remove the water that remained in equilibrium with the oil. Posteriorly, the used oil is submitted to a thermal treatment at temperatures close to 340°C that aim to promote the degradation of the additives used in the lubricating oil formulation and that no chemical degradation during the lubricating life cycle.

In the following step, the used lubricating oil is directed to a reactor where sulfuric acid is added and, through sulfonation reactions, the thermally and chemically degraded products are separated from the base oil.

The oil is sent to a new decanter where is performed the separation of base oil, which is in a liquid phase and the acid sludge (solid phase) that contains the degraded products during the lubricating life cycle.

In the clarifying step, activated clay is added in the oil to remove compounds that can produce color to the base oil. In this step, the acidity of the base oil is controlled through calcium oxide (CaO) addition. Next, the oil passes through distillation steps to remove the lighter compounds that can be produced by thermal cracking during the process or that contaminate the used lubricating oil during the life cycle.

The final process step is base oil filtration to separate the clay that is directed to recovery in the process. Then, the base lubricating oil is pumped to storage tanks.

Due to its simplicity, the acid-clay process needs relatively low capital investment. However, due to the high quantities of clay and sulfuric acid demanded, the operational cost is very high. The main acid-clay process disadvantage is acid-sludge production, which is a hazardous residue with a very difficult treatment.

Another great disadvantage of the acid-clay process is that the technology can produce only Group I base oils. This fact limits the consumer market and strongly impacts profitability when compared with other available technologies.

Another process technology widely employed in the used lubricating oil re-refining is the process called wiped film evaporator. In this process, the used lubricating oil passes through a deasphalting step under vacuum, as described in Figure 6.6.

After the feed stream dehydration step, the lighter fractions are separated in flash vessels. The heavy fraction is heated and sent to wiped film evaporators where the deasphalting process under vacuum at temperatures close to 350°C occurs. The

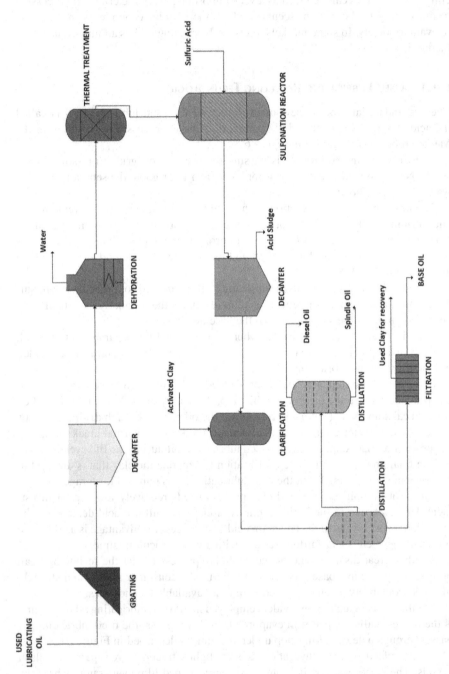

FIGURE 6.5 Basic Process Flow Diagram for the Acid-Clay Process (Meiken Process)

Lubricating Production Refineries

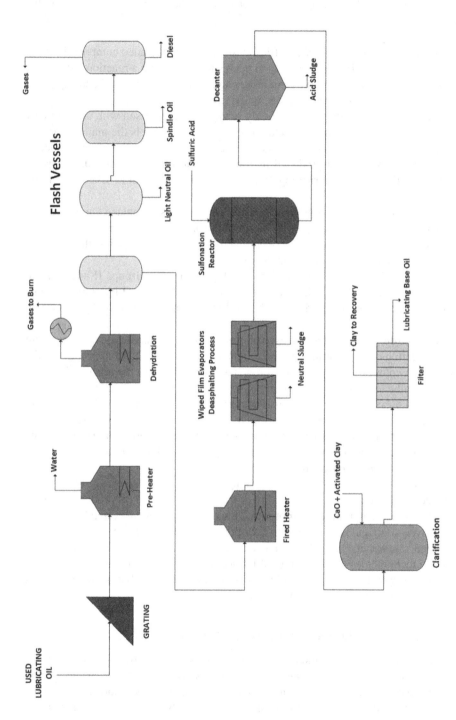

FIGURE 6.6 Process Flow Diagram for Wiped Film Re-refining Process

heavier fraction of the oil containing a major part of the contaminants and degraded lubricating products is removed like neutral sludge, which can be directed to the asphalt industry.

The deasphalting process is insufficient to remove all degraded compounds of the used lubricating oil. Hence, the used oil passes for a step with a reaction with sulfuric acid (sulfonation), where the remaining contaminants are removed and separated from the base lube oil like acid sludge in a decanter.

The acid sludge production is strongly reduced in comparison with the acid-clay process, and the environmental impact is quite reduced. This is the main advantage of the wiped film evaporator process when compared with the acid-clay process. Nevertheless, the base oil produced by this route attends only to the quality specifications of the Group I lubricating oils, according to American Petroleum Institute (API) classification.

Some re-refiners apply a propane deasphalting process to replace the wiped film evaporators. However, the necessity of constant solvent replacement and additional security requirements can raise the operational costs in this case, although the capital investment is similar for both technologies.

The base lubricating oil production with higher quality (Groups II and III) can be reached through the application of hydrotreating processes. Figure 6.7 presents a basic process flow for used lubricating oil re-refining through the hydrorefining route.

After the dehydration step, the used oil passes through a distillation column where the lubricating oil is separated from the lighter fractions. The bottom product of the distillation column is mixed with hydrogen and heated before the reactor. The process is conducted under mild conditions with temperatures varying from 250°C to 300°C and pressures close to 30 bar. The normally applied catalyst is $Co-Mo/Al_2O_3$.

After the reactor, the mixture is sent to a separation drum where the hydrogen is separated from the oil phase and sent to a gas scrubber and then recycled to the reactor. The oil phase is sent to a vacuum tower, where the lubricating oil is separated from water and light compounds.

The main advantages of the hydrorefining route in comparison to the other available technologies are the higher quality of the base lubricating oil produced and the reduction in the environmental footprint. However, the necessary capital investment is relatively high and is only attractive for process plants with large capacity.

Some researchers have dedicated their efforts to the development of new re-refining technologies for treating used lubricating oils, and some of these technologies have shown promising, like ultrafiltration in membranes, but the technology is still in an initial stage of development.

6.2 A GLANCE OVER THE BRAZILIAN LUBRICATING MARKET

The Brazilian domestic market of paraffinic oils is supplied by two refineries that apply the solvent route to produce lubricating oils and waxes for a variety of consumers like food and cosmetic industries, among others. The national lubricating production in 2019 was 3,5 million barrels. Additionally, the internal market is also

Lubricating Production Refineries

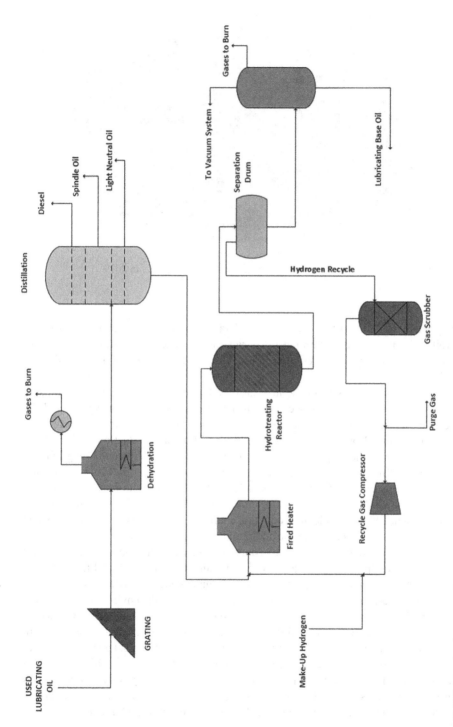

FIGURE 6.7 Basic Arrangement to Hydrotreating Process for Re-refining of Used Lubricating Oil

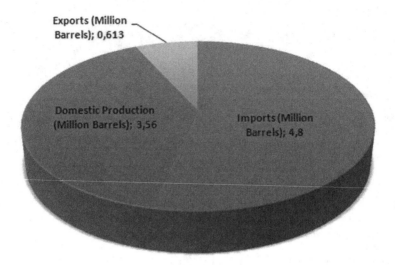

FIGURE 6.8 Balance of the Brazilian Market for Lubricating Oils in 2019

supplied by some importers. According to data from the Brazilian Petroleum Agency (ANP), the internal consumption of lubricating oil reached 7,7 million barrels in 2019. Figure 6.8 shows the composition of the Brazilian market for lubricating oil in 2019.

In Brazil, close to 40% of the lubricating oil consumption is recovered and sent to processing by re-refiners, according to the Brazilian Petroleum Agency (ANP). Despite this significant data, the Brazilian production of lubricating oils is focused on Group I and II oils. As aforementioned, the Brazilian production is carried out by two refineries that apply the solvent route as there aren't hydrocracking units in operation in Brazil. A third Brazilian refinery is capable of producing naphthenic lubricating oils for specific markets as isolators for electrical equipment.

As aforementioned, the re-refining market contributes to reducing the environmental impact of the petroleum production chain as it eliminates a hazardous residue and minimizes the necessity of the additional crude oil quantities.

The necessity to minimize the environmental impact of the industrial activity and economic development is increasingly relevant and a legitimate demand from society, and being very important and having a high environmental impact, the crude oil processing chain needs to adapt its processes to minimize the environmental impact. The circular economy concept is not a new topic, but never in history have we faced a necessity to discuss effective actions like now. Despite being a less discussed item in relation to plastic recycling, the used lubricating oils present a great damage potential to the environment, and recycling seems the best route to remove this waste from nature, minimizing the consumption of new resources in a profitable manner. As discussed previously, recycling used lubricating oil may have a fundamental role in the lubricating market according to the domestic scenario of each refiner.

BIBLIOGRAPHY

1. Audibert, F. *Waste Engine Oils—Rerefining and Energy Recovery*. 1st edition, Elsevier, 2006.
2. Mckinsey & Company. *Lubes Growth Opportunities Remain Despite Switch to Electric Vehicles*, Mckinsey & Company, 2018.
3. Zhu, F., Hoehn, R., Thakkar, V., Yuh, E. *Hydroprocessing for Clean Energy—Design, Operation, and Optimization*. 1st edition, Wiley Press, 2017.
4. Energy Research Company (EPE). *Prospects for the Implementation of Small Refineries in the Brazil*. Technical Note 01/2019.
5. National Agency of Petroleum, Natural Gas and Byofuels (ANP). *Fuel Production and Supply Opportunities in Brazil*, ANP, 2017.
6. Energy International Agency (EIA). *Country Analysis Brief: Brazil*, EIA, 2017.
7. Gran View Research. Lubricants Market Size, Share & Trends Analysis Report by Product (Industrial, Automotive, Marine, Aerospace), By Region, And Segment Forecast, 2019–2025, 2019.

7 Refining Configurations

As with any industrial activity, the crude oil refining industry aims to get profits through the commercialization of derivatives of interest to society. In this sense, the downstream sector aims to add value to crude oil through a series of chemical and physical processes aiming to obtain marketable crude oil derivatives with as low an environmental impact as possible.

Refiners' profitability is directly proportional to their capacity to add value to the processed crude oil to maximize the production of high-added-value streams and derivatives. Equation 7.1 presents a simplified concept of the liquid refining margin.

$$Liquid\ Refining\ Margin = \sum\nolimits_{i}^{n}\left(Di \times vi\right) - Pc - \left(Fc + Vc\right) \quad (7.1)$$

The first term in Equation 7.1 corresponds to the obtained revenue through the commercialization of crude oil derivatives, represented by the sum of the product between the derivative market value and the volume or weight commercialized. As aforementioned, the profitability or refining margin is directly proportional to the refinery capacity to add value to the processed crude slate. The maximization of higher-added-value derivatives leads to the maximization of the first term in Equation 7.1.

The players capable of maximizing the yield of higher-added-value crude oil derivatives in the refining hardware reach a highlighted competitive advantage in the downstream market, and this is the main driving force in the choice of a refining scheme.

Refineries' conception or refining scheme adopted by refiners depends on the market that will be attended aim to define what derivative will be maximized (diesel, gasoline, lubricants, etc.), as well as the quality and environmental requirements that these derivatives need to meet and, of course, crude oil, which will be processed. As generally known, heavier crude oil needs higher conversion and treating levels, raising the processing costs.

The comparison between different refineries is a hard task given that each operational unit attends to distinct markets and different specifications. However, some standard refining schemes were defined over the years in the sense of allowing comparative studies. A refining scheme is the sequence of processing units through which crude oil is submitted to produce desired derivatives that meet the quality and environmental requirements.

The crude oil refining scheme considered basic is called topping. In this case, only a separation process such as atmospheric distillation is applied. Figure 7.1 presents a basic process flow to a typical topping refinery.

Nowadays, this refining scheme is impracticable once purely physical processes are applied and difficultly achieve crude oil conversions sufficiently attractive economically. Furthermore, the derivatives produced have high contaminant content,

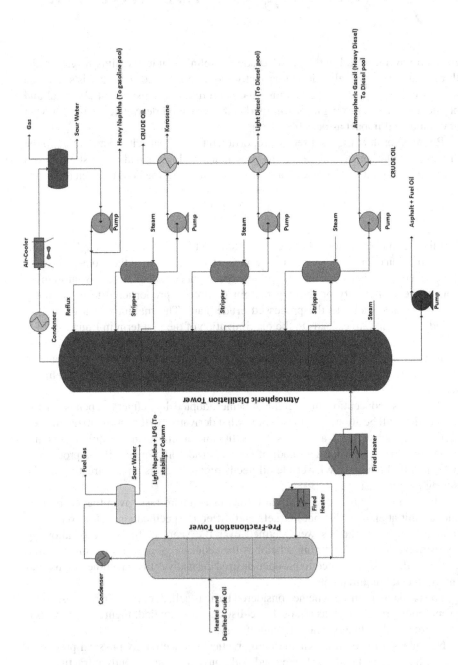

FIGURE 7.1 Typical Scheme for a Topping Refinery

Refining Configurations

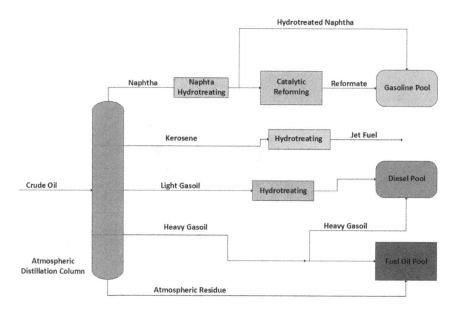

FIGURE 7.2 Block Diagram for a Typical Hydroskimming Refinery

mainly nitrogen and sulfur, making the commercialization of these products prohibitive without breaking current environmental regulations.

Another problem with this refining configuration is the large quantity of asphalt and fuel oil produced. These products have low added value and a restricted market. For these reasons, the topping refining scheme makes the refinery poor and economically uncompetitive. It's uncommon to find a topping refinery currently.

The hydroskimming configuration aggregates conversion and treating processes to the refining scheme, making the refinery operation more profitable and raising the derivative quality. Figure 7.2 shows a block diagram for a typical refinery operating under hydroskimming configuration.

The inclusion of conversion units as catalytic reforming and treating like hydrotreatment raises the derivatives' quality and makes these products more friendly to the environment, improving the added value and allowing their commercialization according to current environmental regulations. However, the hydroskimming configuration still shows a limited conversion and a large production of low-added-value products like fuel oil and asphalt.

Cracking configuration adds to the refining scheme processing units capable of raising the derivatives' recovery from crude oil and units capable of converting residual streams into high-quality derivatives. A cracking refinery has, beyond the units of hydroskimming configuration, vacuum distillation unit, FCC, alkylation unit, visbreaking (nowadays, deasphalting units are more common), and MTBE production process (nowadays in disuse too); the last was applied to raise the octane number of the final gasoline.

Figure 7.3 presents a block diagram for a typical refinery operating under a cracking refining configuration.

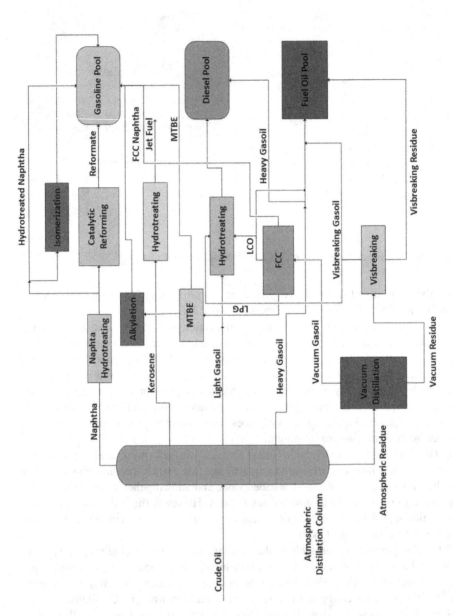

FIGURE 7.3 Typical Process Arrangement to a Cracking Refinery Configuration

Refining Configurations

The vacuum distillation unit increases derivatives' recovery from crude oil. The gas oil produced is fed to an FCC unit producing cracked naphtha that is incorporated into the gasoline pool. Currently, this stream passes through a specific hydrotreatment unit (cracked naphtha hydrotreating unit) to control the contaminant content in the automotive gasoline.

Vacuum residue is fed to the visbreaking unit. As aforementioned, in the modern refining schemes, the installation of solvent deasphalting units is more common. LPG fractions from FCC units can be sent to catalytic alkylation and MTBE production units to produce streams capable of raising the octane number of the final produced gasoline.

Diesel production is also elevated in this case through the addition of light cycle oil (LCO) stream into the diesel pool. The gas oil produced in the visbreaking unit is also added to the diesel pool. However, currently, it's necessary to treat these streams before adding them the diesel pool.

Despite representing a great evolution when compared with the hydroskimming configuration, the cracking refining scheme still conducts production of a large amount of fuel oil. To reduce the production of low-added-value derivatives, it's necessary to install bottom barrel conversion units capable of destroying residual streams and converting them into light and middle derivatives. The coking/hydrocracking refining configurations present these characteristics, as presented in Figure 7.4.

In the case of the coking/hydrocracking refining scheme, fuel oil production is reduced to the minimum necessary to attend to the consumer market. Delayed coking and hydrocracking units raise the production of high-added-value products, like naphtha, diesel, and jet fuel, leading to a significant rise in the refiner profitability.

The improvement in the refinery conversion grade raises the complexity of the refining scheme and, despite the improved profitability, operational costs are also higher in more complex refineries. However, the higher volume and better quality of the produced derivatives produce sufficient elevation in the refining margin to cover these additional costs.

Refineries considered high conversion can include in the refining scheme gasification units to consume the coke produced in delayed coking units, these units can be associated with the power generation unit, and an example is the Flexicoking technology, licensed by ExxonMobil.

There are variations of the presented refining configurations. An example is the combination of FCC and hydrocracking in the same refining scheme. In high conversion units, it's possible to send the unconverted residue from an FCC to a hydrocracking unit. Refining schemes that combine catalytic cracking and delayed cocking technologies are also common. The refining schemes shown in this technical note are optimized to produce automotive fuels. Some refiners direct part of the derivatives to the petrochemical intermediate market due to the high profitability and necessity to attend to a specific consumer market.

As aforementioned, the choice of the adequate refining scheme depends on the assumptions that were adopted in the refinery concept step, the main production focus (fuels or lubricants, for example), and the consumer market that will be attended. More complex refining schemes improve the refinery competitiveness as

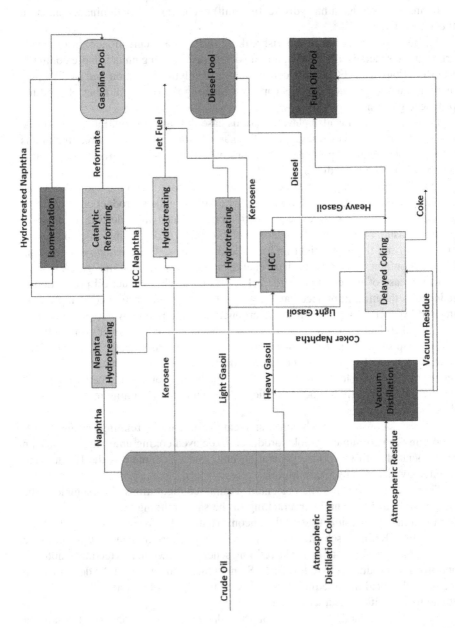

FIGURE 7.4 Process Arrangement to a Refinery Operating under Coking/Hydrocracking Configuration

TABLE 7.1
Complexity Index for Crude Oil Refining Processes

Refining Process	Complexity Index
Atmospheric distillation	1
Vacuum distillation	2
Delayed coking	6
Catalytic cracking	6
Catalytic reforming	5
Hydrocracking	6
Alkylation	10
Aromatics/isomerization	15
Lubricating production	10
Oxygenates production	10

Source: J.H. Gary, G.E. Handwerk, and M.J. Kaiser, *Petroleum Refining: Technology and Economics*, 5th edition, CRC Press, 2007

they are more flexible in relation to crude oil that will be processed. These are strategic characteristics of the current scenario of the refining industry.

7.1 NELSON COMPLEXITY INDEX

The Nelson complexity index is a factor created to ensure a comparative base among the refining hardware. The principle of the Nelson index is the comparison of the conversion capacity with the simple distillation capacity in refining hardware or crude oil refinery.

Most complex refineries can maximize the added value to crude oil through conversion processes like FCC, hydrocracking, and so on, which are capable of improving the yield of higher-added-value derivatives, like gasoline, kerosene, and diesel, through molecular management and residue upgrading processes. The Nelson index is calculated by applying established complexity indexes as presented in Table 7.1.

Considering the complexity indexes presented in Table 7.1, the Nelson index is calculated through Equation 7.2:

$$NCI = \sum_{i=1}^{N} Fi * \frac{Ci}{CD} \quad (7.2)$$

Where Fi = complexity index of the processing unit, Ci = capacity of the processing unit, and CD = capacity of the crude distillation unit

7.2 THE EFFECT OF CRUDE OIL SLATE OVER THE REFINING SCHEME

The continuous supply of adequate crude oil to the refining hardware is one of the assumptions adopted by the refiners for the installation of refining assets or economic analysis of already installed units. However, according to the installed geopolitical scenario, the supply of adequate crude oil to the refining hardware can be seriously threatened, mainly to refiners that operate with lighter and high-cost crudes.

In this sense, more flexible refining hardware in relation to the processed crude slate is an important competitive advantage in the downstream sector, mainly the processing of heavy and extra-heavy crudes due to their lower acquisition cost when compared with the lighter crude oil. The difference in the acquisition cost between these oils is based on the yield of high-added-value streams, which these oils present in the distillation process. Once the lighter crudes normally show higher yields of distillates than the heavier crudes, their market value tends to be higher.

The processing of heavy crudes shows some technological challenges to refiners as, due to their lower yield in distillates, it's necessary to install deep conversion technologies to produce added-value streams that meet the current quality and environmental requirements. Furthermore, the concentration of contaminants, like metals, nitrogen, sulfur, and residual carbon, tends to be high in the heavier crudes, making the processing of the intermediate streams even more challenging.

Despite this fact, the processing of lighter crude slates can lead to issues with the refining hardware, and it is difficult to meet the specifications of some crude oil derivatives, mainly the requirements related to the cold start characteristics of middle distillates. In both cases, an adequate study needs to be considered by the refiners before the decision to process different crude oil slates in the refining hardware considering the profitability and reliability issues in the short and middle terms.

7.2.1 HEAVIER CRUDE OIL PROCESSING

The challenge in the processing of heavy crude oil starts in the desalting step before it is sent to the distillation unit. The desalting process consists basically of the water added to the crude to promote the salt removal from the oil phase that tends to concentrate in the water phase. Figure 7.5 presents a simplified process flow of a crude oil desalting process with two separation stages.

The separation of oil and water phases in the separation drums occurs through the sedimentation process due to the density gap between the water and crude oil. Considering that the sedimentation process can be theoretically described by the Stokes law, according to Equation 7.3:

$$v_R = \frac{1}{18} \frac{d_p^2 g (\rho_p - \rho_f)}{\mu_f} \qquad (7.3)$$

According to Equation 7.3, the sedimentation velocity is proportional to the density gap. In the case of heavier crudes, this gap is lower, leading to a lower sedimentation velocity and the need for higher residence times for adequate separation.

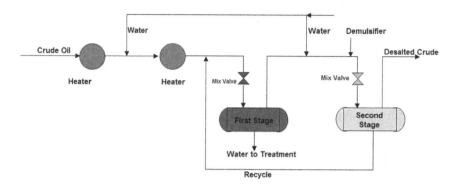

FIGURE 7.5 Crude Oil Desalting Process with Two Separation Stages

TABLE 7.2
Crude Oil Classification Based on API Grade

Classification	API Grade
Light crude	API >31,1
Medium crude	22,3 > API < 31,1
Heavy crude	10,0 > API < 22,3
Extra-heavy crude	API < 10,0

Source: Adapted from Guidelines for Application of the Petroleum Resources Management System, 2011

Another complicating factor in the case of heavy crudes is the higher viscosity of the oil phase that hinders the mass transfer in this phase. Due to these factors, refining hardware designed to process heavy crude oil needs more robust desalting sections taking into account the trend of higher salt concentration in the crude and a harder separation process. Table 7.2 presents an example of crude oil classification based on the API grade.

After the desalting process, crude oil is sent to the atmospheric distillation tower, according to presented in Figure 7.6.

To heavier crudes, the yield of distillates by simple distillation is relatively reduced, and the bottom section in the atmospheric distillation units tends to be overloaded. Table 7.3 presents a comparative analysis of the yields of different crude oils.

Normally, heavy crude oil has a higher concentration of metals, sulfur, and nitrogen. These contaminants tend to be distributed in the intermediate streams concentrating in the heavier streams, making necessary more robust conversion processes and tolerant to these contaminants.

To avoid damage to the catalysts of deep conversion processes such as FCC and hydrocracking, normally refineries that process heavier crudes promote a better fractionating of bottom streams of the vacuum distillation tower. When crude oil presents

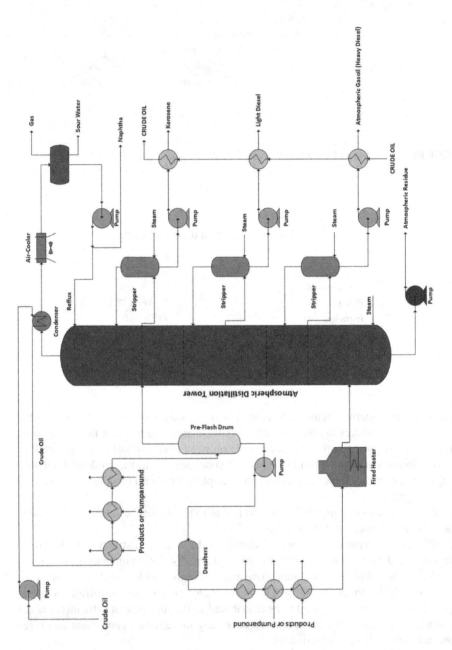

FIGURE 7.6 Atmospheric Distillation Process of Crude Oil

TABLE 7.3
Estimated Yields in the Atmospheric Distillation Process for Different Crude Oils

Crude Oil Type	Naphtha Yield (wt%)	Middle Distillate Yield (wt%)	VGO Yield (wt%)	Vacuum Residue (wt%)	Sulfur Content (wt%)	Specific Gravity
A (API 38,3)	30,1	25,0	30,3	12,3	0,37	0,8333
B (API 35,4)	27,7	34,1	31,2	5,5	0,14	0,8478
C (API 30,1)	16,9	22,6	32,9	26,1	2,00	0,8752
D (API 24,6)	15,8	18,0	32,2	32,9	3,90	0,9065

Where VGO = vacuum gas oil
Source: Modified from Robinson and Hsu, 2019, with Permission

high metal content, it's possible to include a withdrawal of fraction heavier than the heavy gas oil, called residual gas oil or slop cut. This additional cut concentrates the metals in this stream and reduces the residual carbon in the heavy gas oil, minimizing the deactivation process of the conversion processes catalysts as aforementioned. Normally, the residual gas oil is applied as the diluent to produce asphalt or fuel oil.

Due to the high asphaltene content in the heavier crudes, the residual carbon in the bottom barrel streams is also higher than observed in the lighter crudes. This characteristic reinforces even more the necessity of installation processing units with high conversion capacity.

Available technologies for processing bottom barrel streams involve processes that aim to raise the H/C relation in the molecule, either through reducing the carbon quantity (processes based on carbon rejection) or through hydrogen addition. Technologies that involve hydrogen addition encompass hydrotreating and hydrocracking processes, while technologies based on carbon rejection refer to thermal cracking processes like visbreaking, delayed coking, fluid coking, catalytic cracking processes like FCC, and physical separation processes like solvent deasphalting units.

Due to the high content of contaminants in crude oil and, consequently, in the intermediary chains, refining equipment destined for the processing of heavy crudes requires a high hydrotreatment capacity. Usually, the feed streams of deep conversion units like FCC and hydrocracking go through hydrotreatment processes aiming to reduce the sulfur and nitrogen contents and the content of metals. Higher metal and asphaltene content leads to a quick deactivation of the catalysts through high coke deposition rate, catalytic matrix degradation by metals like nickel and vanadium, or even the plugging of catalyst pores produced by the adsorption of metals and high-molecular-weight molecules in the catalyst surface. For this reason, according to the content of asphaltenes and metals in the feed stream, more versatile technologies are adopted to ensure an adequate operational campaign and an effective treatment.

Figure 7.7 presents a scheme of reactants and product flows involved in a heterogeneous catalytic reaction as carried out in hydroprocessing treatments.

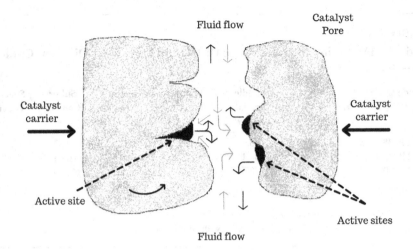

FIGURE 7.7 Reactants and Products Flows in a Generic Porous Catalyst

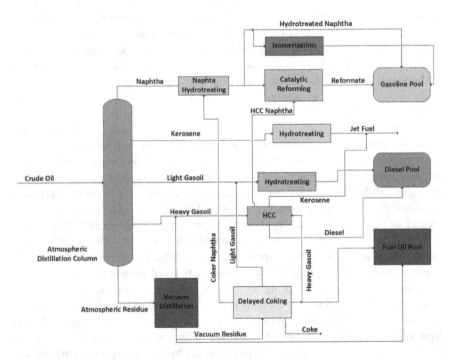

FIGURE 7.8 Process Arrangement to a Refinery Operating under Coking/Hydrocracking Configuration

To carry out hydroprocessing reactions, the mass transfer of reactants to the catalyst pores, adsorption on the active sites to posterior chemical reactions, and desorption are necessary. In the case of bottom barrel stream processing, the high molecular weight and high contaminant content require a higher catalyst porosity to allow the access of these reactants to the active sites, allowing hydrodemetallization, hydrodesulfurization, hydrodenitrogenation, and so on. Furthermore, part of the feed stream can be in the liquid phase, creating additional difficulties to the mass transfer due to the lower diffusivity. To minimize the plugging effect, in fixed-bed reactors, the first beds are filled with higher-porosity solids without catalytic activity and act as filters to the solids present in the feed stream, protecting the most active catalyst from the deactivation (guard beds).

Due to the higher severity and robustness of the processes, the installation cost of the refining hardware capable of processing heavier crudes tends to be higher when compared with the light and medium oils, as well as the operating costs. Figure 7.8 shows a possible refining configuration to be adopted by refiners to add value to heavy crudes.

The refining scheme presented in Figure 7.8, normally called coking/hydrocracking configuration, can ensure high conversion capacity, even with extra-heavy crudes. The presence of hydrocracking units gives great flexibility to the refiner, raising the yield of middle distillates. Figure 7.9 presents a basic process flow diagram for a typical hydrocracking unit designed to process bottom barrel streams.

This configuration is adopted when the contaminant content (especially nitrogen) is high. In this case, the catalyst deactivation is minimized through the reduction of NH_3 and H_2S concentration in the reactors. Among the main hydrocracking process technologies available commercially, we can quote the H-Oil process (by Axens), the EST process (by Eni), the Uniflex process (by UOP), and the LC-Fining technology (by Chevron).

Although the higher processing cost, processing heavy crudes can present a high refining margin. As described earlier, the reduced acquisition cost in relation to lighter crudes, as well as the ease of access and reliability of supply, can make the heavy crudes economically attractive, mainly in countries like Canada, Venezuela, and Mexico that have great reserves of heavy and extra-heavy crude oils.

The flexibility of the refining hardware is a fundamental factor in ensuring the competitiveness of the refiner in the refining market. Normally, the refineries are designed to process a range of crudes, and the wider the range, according to the technical limitations, the more flexible the refinery is related to the processed crude slate. This characteristic is relevant and strategic considering the possibility of enjoying the processing of low-cost crude oil by opportunity besides giving more resilience to a refiner in scenarios of restricting access to the petroleum market, mainly facing a geopolitical crisis.

The current scenario of the downstream industry indicates the tendency for a reduction in transportation fuel demand and an increase in petrochemical intermediate demand, creating the necessity of growing the conversion capacity by the refiners

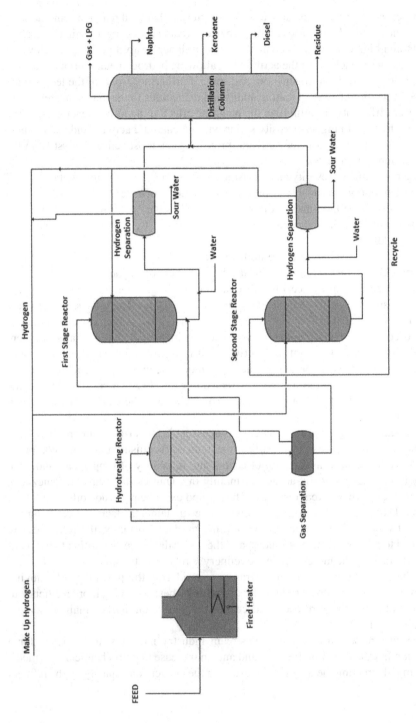

FIGURE 7.9 Typical Hydrocracking Unit Dedicated to Treat Bottom Barrel Streams

in the sense of raising the yield of light olefins in the refining hardware. Furthermore, the new regulation over the marine fuel oil (bunker), IMO 2020, should create even more pressure on the refiners with reduced conversion capacity.

At first, aiming to comply with the new bunker specification, noblest streams, normally directed to middle distillates, should be applied to produce fuel oil with low sulfur content, which should lead to a shortage of intermediate streams to produce these derivatives, raising the prices of these commodities. The market for high-sulfur-content fuel oil should strongly be reduced due to the higher price gap when compared with diesel. Its production will be economically unattractive, leading refiners with low conversion capacity to opt to carry out larger capital investments to give their refining hardware more robustness for the processing of heavier crudes.

The market value of crude oil with higher sulfur content, normally the heavier crudes, tends to reduce after 2020. In this case, refiners with refining hardware capable of adding value to these crudes can have a great competitive advantage in relation to the other refiners, considering the lower acquisition cost of crude oil and higher market value of the derivatives, raising then the refining margins.

As briefly described, heavy crude processing offers technological challenges to refiners. However, according to the geopolitical and the downstream industry scenarios, processing heavier oils can be a competitive advantage. The current scenario of the refining industry indicates a strong tendency to add value through the production of lighter products, mainly petrochemical intermediates. This fact, coupled with the need to produce bottom streams with lower contaminants after 2020 (IMO 2020), increases the pressure even more on refineries with low bottom barrel conversion capacity under the risk of the loss of competitiveness in the market. In this scenario, it is possible to have a strong tendency of resumption in the capital investments in the preparation of these refiners for the processing of petroleum residues and heavier crudes.

7.2.2 Light Crude Oil Processing

The high supply of light oils, especially in the North American market, associated with the tendency of a growing demand for petrochemical intermediates, to the detriment of transportation fuels, has led refiners to look closer integration with petrochemical assets and optimize the refining hardware to increase the yield of petrochemical intermediates.

Despite having a higher yield in derivatives with higher added value, the processing of light oils can offer technological challenges for refiners demanding adjustments in the refining hardware to maintain operational reliability and profitability in view of the higher cost of these raw materials. The high paraffin content, normally present in light oils, tends to favor the precipitation of asphaltenes leading to the need for greater control of crude oil blending to be chosen to compose the crude feedstock of the refinery.

Asphaltene precipitation can also lead to higher fouling rates of the heat exchangers in the preheating network of the atmospheric distillation unit, leading to shorter campaign periods and higher operating costs. In addition, asphaltene precipitation tends to stabilize emulsions, reducing the efficiency of the oil desalting process and leading to the risk of severe corrosion in the top system of the atmospheric distillation column.

The high paraffin concentration of light oils favors the low octane number of direct distillation naphtha and difficulties in specifying the cold start requirements in derivatives such as diesel and jet fuel. In these cases, sections containing dewaxing catalysts can be adopted in the catalytic beds of hydrotreating units, as shown in Figure 7.10.

From the point of view of deep conversion units such as FCC and hydrocracking, the processing of light oils tends to prolong the catalyst life cycle. However, according to the characteristics of the processed oil, some additional care must be taken. Some refiners have reported problems related to the contamination of FCC catalysts by iron, sodium, calcium, and potassium during the processing of American shale oil, leading to difficulties in the fluidization process of the catalyst, increased Sox emissions, and loss of catalytic activity.

In the case of catalytic cracking units (FCC), another point of attention during the processing of light oils is the low availability of coke to maintain the thermal balance of the unit. On the other hand, the feedstock characteristics lead to a high conversion with a tendency to low catalyst consumption and greater olefin production.

The petrochemical industry has been growing at considerably higher rates when compared with the transportation fuel market in the last few years. Additionally, the industry represents a noble destiny and is less environmentally aggressive than crude oil derivatives. The technological bases of the refining and petrochemical industries are similar, which leads to possibilities of synergies capable of reducing operational costs and add value to derivatives produced in the refineries.

During the processing of lighter crude oil, the bottom barrel units tend to be underutilized during light crude processing due to the low yield of bottom currents in the distillation process. However, despite the low flow rates, light petroleum bottom streams tend to present low levels of contaminants, mainly sulfur, which can be especially attractive considering the current scenario of the marine fuel oil market (bunker) with IMO 2020.

The combination of the great availability of light oils and the growing trend of the petrochemical intermediate market seems to be an extremely positive scenario for refiners. However, taking advantage of such opportunities involves reliable refining hardware and adequate and agile uses of available technologies to achieve better competitive positioning.

The adequate choice of crude oil slate is a key planning issue for any refiner and is strictly related to the profitability and reliability of the refining hardware. As discussed previously, heavy and light crude oils offer opportunities and challenges to refiners, and the decision to processing opportunities crude oil, heavier or lighter, like North American shale oil, requires deep analysis to understand the deleterious effect of these crudes on the refining hardware and produced crude oil derivatives once the apparent opportunity can become a risk to the profitability and reliability in the middle term.

As briefly described, the heavier crude processing presents great technological challenges to the refiners and requires higher capital investment to improve the yield

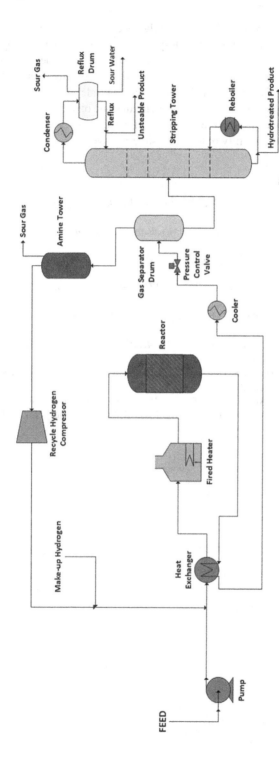

FIGURE 7.10 Typical Arrangement to Hydrotreating Units

of added-value derivatives and, more recently, meet the new specifications of marine fuel oil (bunker) due to IMO 2020. Despite being a less discussed topic, the processing of lighter crude oil presents challenges that need to be considered before the decision to process these crudes in the refining hardware as quoted previously, mainly considering the great availability of lighter crude oil in the current scenario due to the growing production of North American shale oil and the trend of growth of petrochemical participation in the downstream market and its consequent relevance to the economic sustainability of the downstream players.

BIBLIOGRAPHY

1. Fahim, M.A., Al-Sahhaf, T.A., Elkilani, A.S. *Fundamentals of Petroleum Refining*. 1st edition, Elsevier Press, 2010.
2. Gary, J.H., Handwerk, G.E., Kaiser, M.J. *Petroleum Refining: Technology and Economics*. 5th edition, CRC Press, 2007.
3. Robinson, P.R., Hsu, C.S. *Handbook of Petroleum Technology*. 1st edition, Springer, 2017.
4. Speight, J.G. *Heavy and Extra-Heavy Oil Upgrading Technologies*. 1st edition, Elsevier Press, 2013.
5. Ancheyta, J., Speight, J.G. *Hydroprocessing of Heavy Oils and Residua*. 1st edition, CRC Press, 2007.
6. Colombano, A., Colombano, A. *Petroleum Refining & Marketing*. 1st edition, CreateSpace Press, 2017.
7. Robinson, P.R., Hsu, C.S. *Petroleum Science and Technology*. 1st edition, Springer International Publishing, 2019.
8. Guidelines for Application of the Petroleum Resources Management System. November 2011, www.spe.org/industry/docs/PRMS_Guidelines_Nov2011.pdf.

8 Hydrogen Production

The demand for hydrogen increased strongly in the last decades following the necessity of hydrotreatment unit installations in refineries to comply with the pressure to reduce the content of contaminants like sulfur and nitrogen in the petroleum derivates and consequently minimize the environmental impact caused by fuel burn.

Given the greater offer of natural gas in the last few years, the hydrogen generation process through methane (main natural gas component) reforming reactions has consolidated as the principal route to produce hydrogen and syngas to produce the most diversified chemical products like ammonia and convert methanol to olefins (MTO processes).

Regarding hydrogen, this element became a fundamental enabler in the crude oil refining chain. Due to the increasing necessity to reduce the environmental impact of crude oil derivatives, it's practically impossible to produce marketable crude oil derivatives without at least one hydroprocessing step, raising the hydrogen demand as aforementioned. In this sense, even in efforts related to energy transition by the downstream industry, hydrogen presents a key role.

8.1 HYDROGEN AND SYNGAS PRODUCTION ROUTES

Methane steam reforming is currently the most employed technology to produce syngas and hydrogen. Moreover, it's presented until the moment as the most economical route to produce hydrogen on a large scale.

The methane steam reforming process can be chemically represented, as presented here:

$$CH_4 + H_2O = CO + 3H_2 \text{ (Steam Reforming Reaction—Endothermic)}$$
$$CO + H_2O = CO_2 + H_2 \text{ (Shift Reaction—Exothermic)}$$

The reforming reaction is endothermic, so it is favored by higher temperatures (700–850°C), and the catalyst commonly employed is a catalyst with a high content of nickel (Ni) over alumina (Al_2O_3). The reaction equilibrium is favored by lower pressures. However, to avoid the necessity of produced gas compression, the reactions are conducted under moderate pressures (15–25 bar).

Shift reaction is slightly exothermic and occurs under mild reaction conditions (200–350°C) over iron oxide catalyst promoted with cobalt and copper.

Figure 8.1 shows a basic arrangement for the processing unit dedicated to producing hydrogen by methane steam reforming.

Once the nickel catalyst is strongly sensitive to contaminants like sulfur that can cause its deactivation, the process has a treatment step dedicated to removing these contaminants from the methane stream.

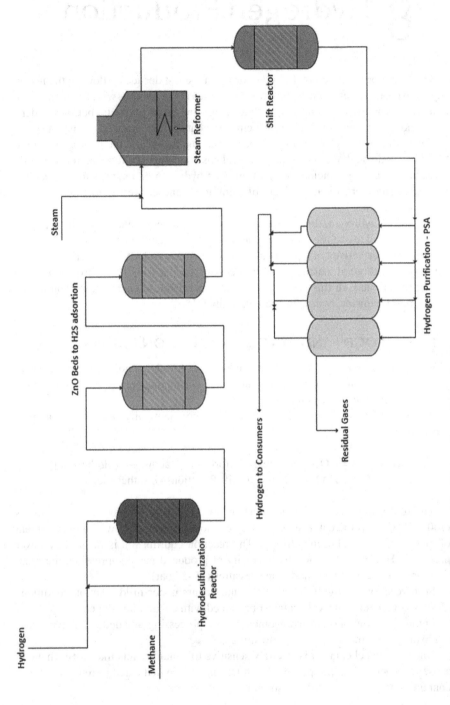

FIGURE 8.1 Basic Process Flow Scheme for Methane Steam Reforming Process

Some undesired reactions can occur during the methane steam reforming process conducting the coke deposition over the catalyst, leading to the loss of chemical activity or a complete deactivation, as described here:

$$2CO = C + CO_2 \text{ (Boudouard Reaction)}$$
$$CH_4 = C + 2H_2 \text{ (Methane Thermal Decomposition)}$$
$$CO + H_2 = C + H_2O \text{ (CO Reduction)}$$

To minimize the risk of carbon deposits, the process is conducted with a higher steam/hydrocarbon ratio (3–4 mole of H_2O per carbon mole). However, the steam/hydrocarbon ratio can't be much high because it can lead to excessive dimensions of the process equipment, and the lower H_2O/hydrocarbon ratio can be compensated by temperature rising. Another side effect of the rise of the H_2O/hydrocarbon ratio is the CO reduction that changes the CO/H_2 ratio. Figure 8.2 presents a basic process flow diagram for a hydrogen generation unit through natural gas steam reforming by Haldor Topsoe.

Another much-studied technology aiming at hydrogen production is called methane dry reforming. The principal chemical reaction of the dry reforming process is presented here:

$$CH_4 + CO_2 = 2CO + 2H_2 \text{ (Methane Dry Reforming Reaction)}$$

The methane dry reforming reaction is endothermic and conducted under high temperatures (higher than 700°C) over a nickel-based catalyst. The dry reforming production route is attractive from the environmental point of view because it can minimize water consumption, and the main reagent is a combustion subproduct that is partially responsible for the greenhouse effect. Another point in favor of dry reforming technology is that the syngas from this process has the ratio $H_2/CO = 1$.

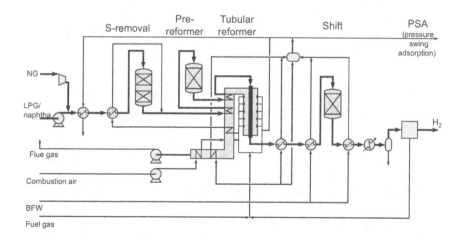

FIGURE 8.2 Hydrogen Generation Process by Haldor Topsoe (with Permission)

This characteristic is ideal for producing oxygenated compounds such as acetic acid and dimethyl ether.

Meanwhile, the main challenge to be overcome in the development of the dry reforming technology is the strong tendency of coke deposition over the catalysts. Due to the nonexistence of water in the process and the low H/C ratio in the process feed, coke formation over the catalysts applied in the dry reforming process is much more severe.

Several researchers have studied ways to develop more resistant catalysts against deactivation by carbon deposition to provide practical application for methane dry reforming technology. Another point to be considered in the dry reforming process development is the necessity of CO_2 purification, which can contribute in a negative way to the economic viability of the process when compared with the steam reforming process.

The methane partial catalytic oxidation process seems attractive as it produces syngas with $H_2/CO = 2$, which is ideal for the Fischer-Tropsch synthesis. The main chemical reaction in the methane partial catalytic oxidation process is presented here:

$$CH_4 + \tfrac{1}{2} O_2 = CO + 2 H_2 \text{ (Partial Oxidation—Exothermic)}$$

This technology applies nickel-based catalysts. However, some research was developed using platinum, palladium, rhodium, and ruthenium as active metals, despite the higher cost of these metals when compared with nickel.

The presence of oxygen in the process reduces the carbon deposit formation over the catalyst, but the use of pure oxygen is not economically competitive because it needs the installation of a cryogenic separation unit to separate the oxygen from the air. On the other hand, using air in the process would increase the equipment dimensions due to the N_2 presence. Another problem associated with methane's partial oxidation is related to the safety owing to the process feed being a mixture of CH_4 and O_2, which can present explosion risk under some process conditions.

Another promising technology to produce syngas and hydrogen from methane is called autothermal reforming process. This technology is the target of several studies and is basically a combination of partial oxidation and steam reforming processes, the principal reactions involved in this process are shown here:

$$CH_4 + 1 \tfrac{1}{2} O_2 = CO + 2 H_2O \text{ (Exothermic)}$$
$$CH_4 + H_2O = CO + 3 H_2 \text{ (Steam Reforming Reaction—Endothermic)}$$
$$CO + H_2O = CO_2 + H_2 \text{ (Shift Reaction—Exothermic)}$$

The process is called autothermal because it involves endothermic and exothermic reactions, and in theory, the heat produced in the partial oxidation reaction is consumed in the reforming reaction.

Autothermal reforming can produce syngas with H_2/CO ratio close to 2, adjusting the reactants proportion in the process feed. Normally the catalyst used in the process is nickel-based catalyst over alumina. However, coke deposition over the catalyst represents a challenge that needs to be overcome in the methane autothermal reforming too.

Hydrogen Production 141

As aforementioned, several technologies have been studied and developed, but at the moment, the methane steam reforming process still is the most economical route to produce hydrogen on a large scale.

8.2 RENEWABLE HYDROGEN GENERATION ROUTES: FUNDAMENTAL ENABLER TO THE ENERGY TRANSITION

In the current scenario, the pressure from society to have energy transition efforts is increasingly high, aiming to reduce fossil fuel participation in the global energetic matrix. As aforementioned, hydroprocessing technologies achieve a fundamental role in any refining hardware, but this fact has a side effect due to the high CO_2 emissions during the natural gas steam reforming process to produce hydrogen.

Some refiners are adopting the co-processing of renewable materials in the crude oil refineries aiming to produce high quality and cleaner transportation fuels. Despite the advantages of environmental footprint reduction in the refining industry operations, renewables processing presents some technological challenges to refiners.

The renewable streams have much unsaturation and oxygen in their molecules, which leads to high heat release rates and high hydrogen consumption. This fact leads to the necessity of a higher capacity of heat removal from hydrotreating reactors to avoid damage to the catalysts. The main chemical reactions associated with the renewable streams hydrotreating process can be represented like so:

$$R\text{-}CH=CH_2 + H_2 \rightarrow R\text{-}CH_2\text{-}CH_3 \text{ (Olefin Saturation)}$$
$$R\text{-}OH + H_2 \rightarrow R\text{-}H + H_2O \text{ (Hydrodeoxigenation)}$$

Where R represents a hydrocarbon

These characteristics lead to the necessity of higher hydrogen production capacity by the refiners and more robust quenching systems of hydrotreating reactors or, in some cases, the reduction of processing capacity to absorb the renewable streams. At this point, it's important to consider a viability analysis related to the use of renewables in the crude oil refineries as the higher necessity of hydrogen generation implies higher CO_2 emissions through the natural gas reforming process, which is the most applied process to produce hydrogen on a commercial scale.

$$CH_4 + H_2O = CO + 3H_2 \text{ (Steam Reforming Reaction—Endothermic)}$$
$$CO + H_2O = CO_2 + H_2 \text{ (Shift Reaction—Exothermic)}$$

This fact leads some technology licensors to dedicate their efforts to looking for alternative routes for hydrogen production on a large scale in a more sustainable manner. Some alternatives pointed out can offer promising advantages:

- Natural gas steam reforming with carbon capture: The carbon capture technology and cost can be a limiting factor among refiners.
- Natural gas steam reforming applying biogas: The main difficulty in this alternative is a reliable source of biogas and its cost.

- Reverse water gas shift reaction ($CO_2 = H_2 + CO$): This is one of the most attractive technologies, mainly to produce renewable syngas.
- Electrolysis: The technology is one of the more promising for the near future.

Nowadays, there is a classification for different hydrogen production routes that is applied according to the level of carbon emissions. This classification is thus summarized:

- Grey hydrogen—hydrogen production from fossil fuels like carbon coal gasification or methane steam reforming
- Blue hydrogen—hydrogen production from fossil fuels but relying on carbon capture technologies to reduce the environmental impact
- Green hydrogen—hydrogen production from renewable raw materials like biomass gasification or electrolysis

As aforementioned, hydrogen is a key enabler to the future of the downstream industry, and the development of renewable sources of hydrogen is fundamental to the success of the efforts to the energy transition to a lower carbon profile.

8.3 HYDROGEN NETWORK AND MANAGEMENT ACTIONS

As mentioned previously, hydrogen became a fundamental production input to modern crude oil refineries, and its adequate management is a key factor in ensuring controlled operating costs and competitiveness in the market and allowing the production of marketable crude oil derivatives. Hydrogen management actions start with a mass balance involving the hydrogen network that is composed of hydrogen sources, hydrogen purification systems, and hydrogen consumers, as presented in Figure 8.3.

Hydrogen generation relies on the refining configuration adopted in the refinery. Normally, refineries that rely on catalytic reforming units apply the hydrogen produced in this processing unit to compose a relevant part of the hydrogen network, becoming an important internal source of hydrogen. As presented previously, the hydrogen generation route most applied in the refining industry is steam reforming based on naphtha or natural gas, as described in Figure 8.1.

Hydrogen purifying technologies are another important part of the hydrogen network. Normally, modern refineries apply pressure swing adsorption (PSA) technologies to purify the hydrogen, reaching purity higher than 99%. Despite this fact, some refiners still use treatments based on amine treatment.

Despite the lower capital cost requirement when compared with PSA technologies, the amine treating units produce hydrogen with low purity, and this represents a great disadvantage, especially to refiners with deep conversion hydroprocessing units. Other hydrogen purifying technologies commercially available are the membrane separations, which can reach a purity of 98%, and the cryogenic processes, which can reach a purity of 96%. Hydrogen purifiers have a key role in hydrogen management as, in controlling hydrogen recovery in off-gases, one of the main sources of hydrogen losses in the refineries is the burning of fuel gas during poor recovery capacity.

Hydrogen Production

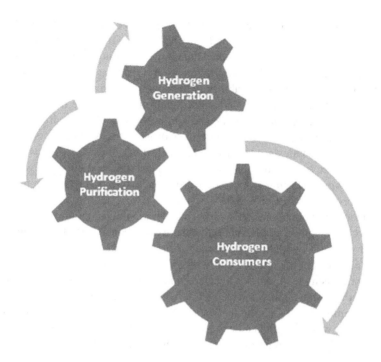

FIGURE 8.3 A Typical Hydrogen Network in a Crude Oil Refinery

According to the refining configuration, hydrogen consumers in a crude oil refinery can vary. An example is a refinery that relies on isomerization units to increase the production of high-quality gasoline.

The high cost of hydrogen generation and the great amount of CO_2 (greenhouse gas) produced are the main driving force for an adequate hydrogen management in the refining hardware. Process integration technologies like the pinch method and mathematical modeling are being applied to reach the most rational use of hydrogen in the refining hardware.

The reliability of hydrogen purification systems and the optimization of hydroprocessing units are fundamental in avoiding the burn of hydrogen in the fuel gas ring or the flare that can raise the operating costs and reduce the refining margins of the refiners. Another key point is the availability of control and instrumentation systems to allow the flow measurement and adequate accuracy of mass balances and actions to define optimization actions and mathematical modeling.

8.3.1 THE ROLE OF CATALYTIC REFORMING UNITS IN THE REFINERIES' HYDROGEN BALANCE

As aforementioned, demand for hydrogen increased strongly in the last decades following the necessity of hydrotreatment unit installations in refineries to comply with the pressure to reduce the content of contaminants like sulfur and nitrogen in the petroleum derivates and consequently minimize the environmental impact caused by fuel burn. This scenario made hydrogen one of the most important production inputs in modern refineries, and adequate hydrogen management actions reach strategic character to keep the operating costs and refining margins under control, contributing to economic sustainability in the downstream industry.

Normally, refineries that rely on catalytic reforming units use the hydrogen produced in this processing unit to compose a relevant part of the hydrogen network, becoming an important internal source of hydrogen. In some markets, where the demand for petrochemicals is lower, the main relevance of catalytic reforming to the refining hardware is the hydrogen generation against the production of light aromatics. Figure 8.4 presents an example of a hydrogen network in a crude oil refinery with high hydroprocessing capacity.

In refineries with bottlenecked hydrogen generation units, the hydrogen from catalytic reforming units is fundamental to ensure compliance with the current quality and environmental regulations, becoming a fundamental enabler to profitable and reliable operations of the refining hardware.

Hydroprocessing technologies became fundamental to the downstream industry both to produce high quality and cleaner derivatives or to prepare feedstocks for the processing units like RFCC, and this dependence increased even more after the start of IMO 2020, which requires a deep treatment of bottom barrel streams to comply with the new quality requirements of the marine fuel oil (bunker). In this sense,

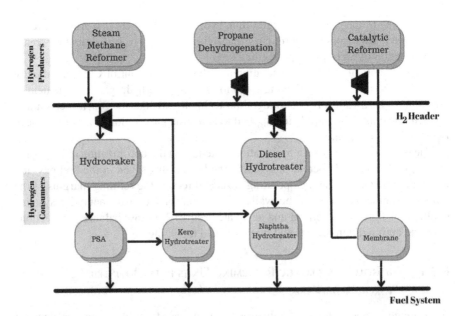

FIGURE 8.4 Example of a Hydrogen Network in a Crude Oil Refinery Relying on Catalytic Reforming

hydrogen generation units achieve strategic character to refiners, and the efficient and reliable operation of these units needs to be a priority to refiners.

Beyond the current status, it's important to understand that the energy transition is no longer a matter of choice for the players of the downstream industry but a reality, and the efforts to find cleaner sources of hydrogen need to be supported by refiners to minimize the environmental impact of the crude oil processing chain while ensuring the production of high-quality and added-value derivatives.

BIBLIOGRAPHY

1. Gary, J.H., Handwerk, G.E., Kaiser, M.J. *Petroleum Refining: Technology and Economics*. 5th edition, CRC Press, 2007.
2. Hilbert, T., Kalyanaraman, M., Novak, B., Gatt, J., Gooding, B., McCarthy, S. *Maximising Premium Distillate by Catalytic Dewaxing*, PTQ Magazine, 2011.
3. Lafleur, A. Use and Optimization of Hydrogen at Oil Refineries. Shell Company, Presented at DOE H_2@Scale Workshop—University of Houston, 2017.
4. Peiretti, A. *Haldor Topsoe—Catalyzing Your Business*, Technical presentation of Haldor Topsoe Company, 2013.
5. Harrison, S.B. *Turquoise Hydrogen Production from Methane Pyrolisis*, PTQ Magazine, 2021.
6. Gupta, K., Aggarwal, I., Ethakota, M. *SMR for Fuel Cell Grade Hydrogen*, PTQ Magazine, 2020.

9 Caustic Treating Processes

Concerns related to the environmental impact of produced derivatives are constant matters to the petroleum refining industry, mainly in the current scenario where society increasingly requires products with lower environmental impact. This appeal by society is translated to severe environmental regulations. At the same time, the downstream industry is demanded to produce derivatives with better performance and quality. This scenario requires the permanent development of the processes dedicated to the treatment of produced derivatives.

The most employed route to achieve low-sulfur crude oil derivatives nowadays is hydroprocessing technologies, especially due to the restricted environmental regulations that require increasingly low contaminant content in the derivatives, especially fuels like gasoline and diesel. Despite this fact, the control of corrosivity and total sulfur content is controlled through caustic treating processes. These treating technologies can also be applied to treating kerosene (jet fuel), but only with intermediate streams from specific crude oils as it is limited to removing sulfur and is incapable of removing nitrogen. Nowadays, the most common route to treat kerosene is hydrotreating technologies.

9.1 CAUSTIC TREATING TECHNOLOGIES

One of the most employed treatment processes is called caustic treating. This process aims to remove sulfur compounds such as H_2S and mercaptans (R-SH), which are responsible for giving corrosivity and odor to the derivatives like LPG and naphtha. Caustic treatment is applied to treat LPG, kerosene, and naphtha fractions, which are incorporated into the gasoline pool. Nowadays, faced with the restrictive sulfur content in the final gasoline, caustic treating has been almost totally substituted by hydroprocessing processes. Beyond sulfur compounds, caustic treatment is capable of removing other detrimental compounds such as nitrogen and naphthenic/carboxylic acids, which gives higher chemical stability to derivatives.

The first caustic treatment process applied was the conventional caustic process. In this case, acid compounds are removed after direct contact with caustic soda solution, according to the following chemical reactions:

$$H_2S + NaOH = Na_2S + H_2O \text{ (}H_2S \text{ Extraction)}$$
$$RSH + NaOH = NaSR + H_2O \text{ (Mercaptan Extraction)}$$

Figure 9.1 presents a process flow diagram for a typical conventional caustic treating unit.

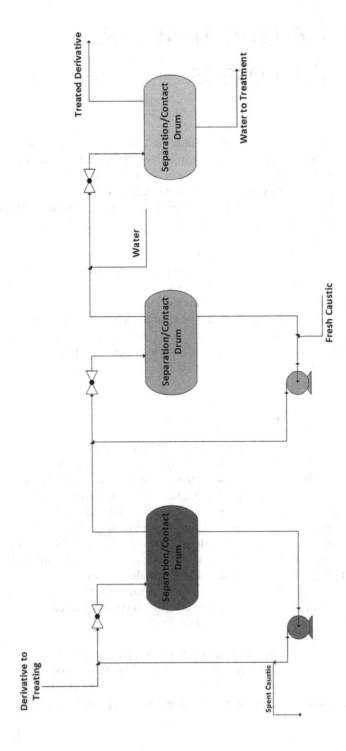

FIGURE 9.1 Conventional Caustic Treatment

Caustic Treating Processes

The feed stream is mixed with the caustic solution through a valve installed at the entrance of the vessels that promote the contact of aqueous and hydrocarbon phases and the posterior separation phases. The vessels are designed to give adequate residence time to allow the extraction of acid compounds from the hydrocarbons. Due to its low specific gravity, the hydrocarbon phase is withdrawn from the top and directed to another vessel, where it is mixed again with the caustic solution. Now with higher concentration, after the second vessel, the hydrocarbon stream receives a water injection and flows to a third vessel that promotes the phase separation. The treated derivative is sent to storage, and the contaminated water is directed to treatment.

The main process variables are the caustic soda content in the caustic solution. Generally, the higher soda content means a higher mercaptan extraction capacity. However, there is an economic limit where the rise in soda concentration promotes an insignificant rise in extraction capacity. Normally, the caustic soda content is controlled through caustic solution density. The other process variables are soda circulation flow rate, the pressure drop in the valves located in the entrance of contact vessels, and operating temperature. Lower temperatures improve the mercaptan extraction. On the other hand, the phase separation is impaired. Normally, the operating temperature is controlled in the range of 35–40°C for conventional caustic processes.

The traditional caustic treating process presents limitations that restrict its use in the modern refining schemes. For feed streams with high H_2S and mercaptan content, the consumption of fresh soda is high, making the process less competitive economically due to high operational costs. Furthermore, there are some environmental restrictions related to the discharge of spent soda. Due to these limitations, the conventional caustic treatment is normally applied to treat feed streams with low contaminant content, such as straight-run products (LPG, naphtha, and kerosene).

Regenerative caustic treatment applies the continuous regeneration of caustic solution. This significantly reduces the soda consumption in the process, leading to a reduction in operational costs and making the caustic process more flexible in relation to the feed stream contaminant concentration. Regenerative caustic treatment can be applied to LPG, naphtha, and kerosene (jet fuel).

The caustic regenerative process is dedicated to removing mercaptans exclusively, so it's necessary to treat the feed stream to remove H_2S. Normally, conventional treating units or amine treatment processes are employed to prepare the feed stream for the regenerative caustic process. The chemical reactions involved in the regenerative caustic treating process are presented as such:

$$RSH + NaOH = NaSR + H_2O \tag{9.1}$$

$$2\,NaSR + H_2O + \tfrac{1}{2}\,O_2 \rightarrow RSSR + 2\,NaOH \tag{9.2}$$

Reaction 9.2 occurs in the presence of a homogeneous cobalt-based catalyst and promotes the caustic solution regeneration applied in the mercaptan extraction. Figure 9.2 presents a process flow diagram of a typical regenerative caustic processing unit dedicated to treating LPG.

150   Crude Oil Refining

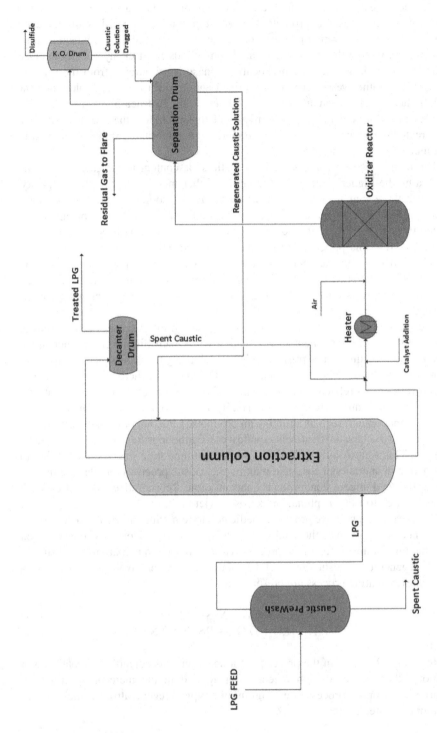

FIGURE 9.2 Typical Process Arrangement to LPG Regenerative Caustic Unit

LPG is fed to a caustic pre-wash drum to remove H_2S and contaminants that can harm the treating process. In the next step, the LPG is directed to an extraction column where occurs the contact between LPG and caustic solution. Treated LPG is recovered at the top of the column, and the spent caustic solution is pumped into the treatment system from the column bottom.

The bottom stream containing the spent caustic solution is heated and receives an air injection before being fed to an oxidizer reactor where the extracted mercaptans are oxidized. After this step, the stream is sent to a separation drum from where the regenerated solution is pumped back to the extraction column, and the disulfides are sent to the refinery fuel gas pool. Nowadays, due to the need to attend to the environmental regulations related to atmospheric emissions, this stream is usually sent to hydrotreating units.

LPG regenerative caustic process promotes sulfur compound extraction. Once the mercaptans are removed in the aqueous phase, in the case of naphtha and jet fuel treatments, the mercaptans are converted to disulfides but kept in the hydrocarbon. In these cases, the regenerative caustic process is called the sweetening process. The total sulfur content in the product does not change, but the derivative becomes less corrosive.

The main process variables of the LPG regenerative caustic process are the temperatures of the extraction column and oxidizer reactor, air and caustic solution flow rates, caustic solution density, and the operating pressure of the extraction column that is controlled in the range to keep the LPG in the liquid phase (13–14 kgf/cm^2).

The main regenerative caustic technology available in the market is the Merox process developed by UOP. However, there are other processes available: the technologies Mericat and Regen, by Merichem, and the processes Sulfrex and Sweetn'K, by Axens.

Jet fuel treating process is capable of meeting the current specifications of maximum sulfur content. However, the removal of nitrogen compounds that are responsible for giving color and chemical instability to the derivative can't be totally carried out through caustic processes. For this reason, in the modern refining schemes, this task is conducted by hydrotreating processes. Refineries that don't have kerosene hydrotreating units normally need to control the nitrogen content in crude oil processed. This leads to an important restriction that can make the refiner less competitive.

Figure 9.3 presents a basic process flow diagram for a typical jet fuel regenerative caustic treating unit. The clay bed is added in the process scheme to remove nitrogen compounds and oil-soluble contaminants.

As aforementioned, the caustic treatment of jet fuel (kerosene) and naphtha has been applied less by the refiners as the environmental regulations and quality specifications of these derivatives are better met through hydrotreatment processes. However, for LPG, the caustic process is still relevant.

A major part of the LPG supplied to the market is produced by residue conversion processes such as FCC and delayed coking. Due to the high contaminant content in these feed streams, the produced LPG presents a high sulfur content and corrosivity. Therefore, the caustic treating processing units are fundamental to meeting the

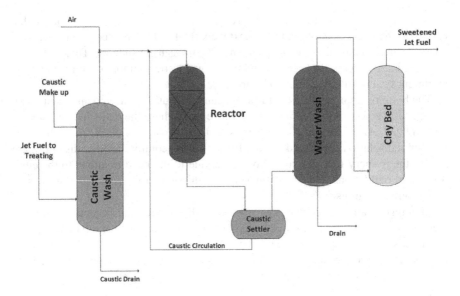

FIGURE 9.3 Typical Flow Scheme for Jet Fuel Regenerative Caustic Treating Process

quality specifications of the derivatives. This ensures the consumer market supply and the refiner's profitability.

9.2 BENDER TREATING TECHNOLOGIES

The bender process aims to reduce the corrosivity of crude oil derivatives, especially kerosene, through the sweetening process, where the corrosive mercaptans are converted to less aggressive disulfides through catalytic oxidation. The main chemical reactions involved in the process are as follows:

$$2\ R\text{-}SH + \tfrac{1}{2}\ O_2 \rightarrow RSSR + H_2O \tag{9.3}$$
$$2\ R\text{-}SH + S + 2\ NaOH \rightarrow RSSR + Na_2S + H_2O \tag{9.4}$$

Reactions 9.3 and 9.4 are carried out in a fixed-bed reactor, applying lead oxide (PbO) as a catalyst, which is converted to lead sulfide (PbS) in the processing unit.

The bender process is not efficient in treating nitrogen compounds, being rarely applied nowadays due to its limitations in meeting the current requirements of crude oil derivatives. As aforementioned, to meet the current specifications of crude oil derivatives, the most efficient route is hydroprocessing technologies.

Minimizing the environmental impact of operations is a constant goal in the crude oil refining industry, and treating technologies capable of minimizing the contaminants in crude oil derivatives is essential to achieve this goal. As quoted, hydroprocessing technologies are the main route to comply with the current quality and environmental requirements for crude oil derivatives as the alternative treating methods present severe limitations, as shown earlier, but caustic treating technologies are still relevant to control the quality of LPG, which has a great demand in some markets.

BIBLIOGRAPHY

1. Fahim, M.A., Al-Sahhaf, T.A., Elkilani, A.S. *Fundamentals of Petroleum Refining*. 1st edition, Elsevier Press, 2010.
2. Gary, J.H., Handwerk, G.E., Kaiser, M.J. *Petroleum Refining: Technology and Economics*. 5th edition, CRC Press, 2007.
3. Moulijn, J.A., Makkee, M., Van-Diepen, A.E. *Chemical Process Technology*. 2nd edition, John Wiley & Sons Ltd., 2013.
4. Coker, A.K. *Petroleum Refining Design and Applications*. 1st edition, John Wiley & Sons Ltd., 2018.
5. Speight, J.G. *Handbook of Petroleum Refining*. 1st edition, CRC Press, 2020.
6. Brouwer, M.P. *Oil Refining and the Petroleum Industry*. 1st edition, New Science Publishers, 2012.

10 Environmental Processes

Just like any other chemical processes industry, oil refining presents a great environmental impact. For decades, engineers, scientists, and researchers have dedicated efforts to minimize the environmental footprint of petroleum refining. Some of the major impacts produced by crude oil processing are water and atmospheric emissions.

To keep the environmental impacts in the crude oil refining sector under control, some processing technologies were developed over the years and installed in the refining hardware. Nowadays, it's impossible to think in the downstream sector without the environmental processes units due to the current environmental requirements, and the performance and reliability of these units are fundamental to the refiners' strategy to achieve the most profitable and cleaner operations.

10.1 SOUR WATER STRIPPING TECHNOLOGIES

The petroleum derivate production needs a large amount of water for cooling fluid, steam generation, or direct use in the process like in the crude oil desalting step. Water has become an increasingly scarce resource, and any effort dedicated to reducing the volume applied in the process is welcome.

One of the most important environmental processing units in a petroleum refinery is the so-called sour water stripping unit. Sour water is the water that had contact with the petroleum or its derivates during some step in the process. This contact can be like rectification steam in distillations columns or in contact with hydrocarbon phases. Contaminants like NH_3 and H_2S tend to concentrate in the aqueous phase, so sour water commonly has a high concentration of these compounds.

The sour water stripping unit uses the concept of fluid rectification with steam and partial pressure reduction to move the phase equilibrium to the vapor phase, releasing the contaminants from the liquid, as presented in Figure 10.1.

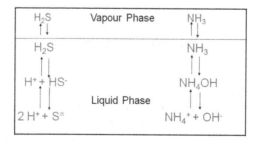

FIGURE 10.1 Phase Equilibrium in the Sour Water Stripping Process

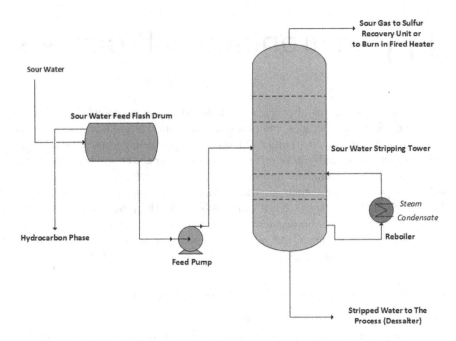

FIGURE 10.2 Typical Arrangement for Sour Water Stripping Unit with a Single Tower

Like any other process technology, the sour water stripping process was developed and improved over time, mainly to reduce atmospheric emissions and to raise the water reuse in the refineries.

The initial design concept for sour water stripping units had one rectifying tower. In this tower, both contaminants (NH_3 and H_2S) are removed and form the stream called sour gas, as described in Figure 10.2.

In these cases, the tower operates with relatively low pressure (about 1 kgf/cm²).

Initially, the designs predict to send the sour gas to burn in fired heaters, like in distillation units. Nowadays, with the environmental restrictions and the necessity to reduce SOx and NOx emissions, the project concept was changed, and the sour gases are directed to sulfur recovery units with a chamber to convert the NH_3 to N_2. This is necessary to avoid that the NH_3 harm the H_2S conversion into sulfur through the Claus process.

The modern designs rely upon the installation of two towers, one for H_2S removal and the second for NH_3 removal, as described in Figure 10.3.

For units with two towers, the H_2S rectifier operates under pressures of about 5–11 kgf/cm², while the ammonia rectifier operates under pressures of about 1–2 kgf/cm². The arrangement with two towers shows some advantages in relation to the project with a single tower, as that allows higher recovery of H_2S like elemental sulfur, reducing the SOx emissions. Furthermore, the design with two towers allows the recovery of ammonia present in the sour water or converts this stream to N_2.

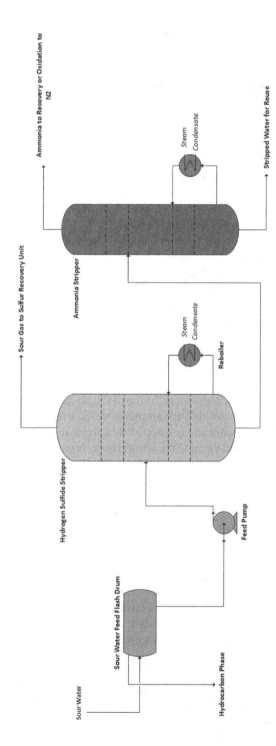

FIGURE 10.3 Typical Arrangement for Sour Water Stripping Unit with Two Towers

As a disadvantage, in comparison with the single-tower design, we can mention the higher initial investment, higher energy consumption, and increased operational complexity.

10.2 AMINE TREATING TECHNOLOGIES

One of the main impacts of the refining industry is the atmospheric emissions due to the burning of fuel gases (natural gas, LPG, and fuel gas) in process equipment like fired heaters and boilers.

Reduction of contaminant content such as H_2S and CO_2 in gaseous streams in the crude refining and gas industries is commonly carried out through the amine treating processing unit. This technology is based on reversible chemical absorption of CO_2 and H_2S in alcohol amines. This process can be applied to treat natural gas, LPG, fuel gases, and recycled gases in hydrotreating units.

In the refining industry, amine treating units are fundamental to meet the corrosivity and H_2S content requirements in the LPG and fuel gas. The process consists of the treatment of a gas stream with an amine solution, contaminant absorption, and amine solution regeneration, as presented in Figure 10.4.

The feed stream is fed in the bottom of the absorption column, where H_2S and CO_2 are removed according to the following chemical reactions:

$$H_2S + R_2NH \leftrightarrow R_2NH_2^+ + HS^- \quad (10.1) \; H_2S \; \text{Absorption}$$
$$CO_2 + 2R_2NH \leftrightarrow R_2NCOO^- + R_2NH_2^+ \quad (10.2) \; CO_2 \; \text{Absorption (Simplified)}$$

The treated gas stream is withdrawn at the top of the absorption column while the amine stream with high H_2S content (rich amine) is sent to a regeneration step where the rich amine is fed to a column and is stripped with steam to remove H_2S and CO_2. The bottom stream is pumped again to the absorption column. This stream is called a lean amine solution. This stream passes through filters to remove particulates and impurities. Before reaching the absorption tower, the lean amine exchanges heat with the rich amine stream.

The stream recovered at the top of the regeneration tower is called a sour gas stream and has a high H_2S content. Normally this stream is directed to sulfur recovery units where the H_2S is converted into elemental sulfur through the Claus process. For units dedicated to treating LPG, there is a feed stream washing step (with water) to remove insoluble compounds (mercaptans, thiocyanide, etc.).

The main process variables of the amine treating process are pressure in the regeneration column (1,5–1,8 bar), the temperature of amine solution in the absorption and regeneration columns, the temperature at the top of regeneration column, H_2S content in the lean amine stream, and the temperature in the bottom of regeneration column (115–125°C).

An important control parameter in the process is the solid content in the lean amine stream. A high solid content can lead to a foaming formation that makes the process control difficult and leads to the drag of the amine solution by the gaseous streams. To avoid solid accumulation, the lean amine stream is filtered through carbon filters (to remove organic contaminants) and cartridge filters (to remove

Environmental Processes

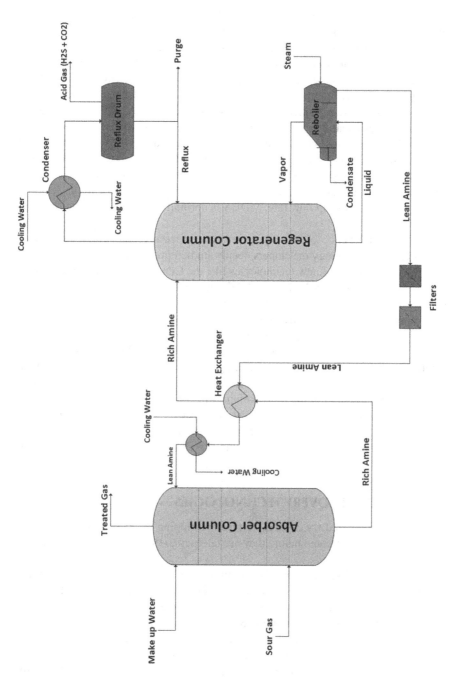

FIGURE 10.4 Basic Process Scheme for a Typical Amine Treating Processing Unit

inorganic solids). Other relevant variables are the amine circulation flow rate and free amine content. In other words, the quantity of amine that doesn't undergo thermal or chemical degradation during the process. Normally, the fixed amine content (that undergoes degradation) is controlled to be below 5%.

The amine type that is applied also has a great influence on the process results. Amines normally employed are MEA (monoethanolamine), DEA (diethanolamine), and MDEA (methyl-diethanolamine). MEA is normally applied in the natural gas treatment once the acid compound content is lower. Due to high chemical activity, the MEA is normally harder to recover when compared with DEA, which is commonly used in refineries' treating units. The MEA is applied in solutions with 15–20% in mass, while DEA is applied in solutions, 20–30% in mass. MDEA presents lower H_2S removing efficiency and normally is not applied to remove CO_2. On the other hand, the lower activity is easily regenerated when compared with MEA and DEA. However, MDEA has a higher cost and a greater solubility in hydrocarbons, leading to losses and high makeup flow rate. MDEA is normally employed in solutions with 40–50% in mass.

Due to the relevance of the amine treating process to the refining industry, researchers and technology developers deeply studied process improvements, which led to the development of new technologies like the Cansolv process, by Shell Company; the process Flexsorb, by ExxonMobil; the Amine Guard technology, by UOP; and the Sultimate process commercialized by Prosernat.

As aforementioned, amine treating of gaseous streams such as natural gas, LPG, and fuel gas is essential to downstream industry to meet the quality and environmental requirements necessary to commercialize these products. Amine treating technologies are also essential to the refining scheme as they act as an intermediate step in processes such as hydrotreating, sulfur recovery (tail gas units), and in some cases, hydrogen generation units. Therefore, the constant search for improvements to existing processes and the development of new technologies capable of reducing the environmental impact of the process is very attractive to the refining industry, which has a principal driving force in environmental impact reduction and raising the refining margin.

10.3 SULFUR RECOVERY TECHNOLOGIES

After the processing of dirty gases in sour water stripping or amine treating units, the sour gases need to be treated before they are released to the atmosphere, minimizing the SOx emissions.

One of the most important processing units in the refining complex is the sulfur recovery unit. This unit is responsible for recovery, in the elemental sulfur form, the sulfur removed from process streams treated in sweetening units of light fractions (LPG and fuel gas) produced in the deep conversion units such as delayed coking, FCC, and hydrotreating. Furthermore, the gaseous streams produced in the sour water stripping unit are directed to the sulfur recovery unit.

The sulfur recovery unit feed stream, called sour gas, is composed basically of H_2S (50–80%) and contaminants like CO_2, H_2O, NH_3, and hydrocarbons. The most employed technology to recover sulfur in the refining industry is the Claus process.

The Claus process is based on two H_2S conversion steps, a thermal step followed by a catalytic step. In the thermal step, the H_2S is partially burned according to the following chemical reaction:

$$H_2S + 3/2\ O_2 \rightarrow SO_2 + H_2O \qquad (10.3)$$

Then the remaining H_2S reacts with the SO_2, producing elemental sulfur according to the following chemical reaction:

$$2H_2S + SO_2 \leftrightarrow 3S + 2H_2O \qquad (10.4)$$

The global Claus process chemical reaction is the following:

$$3H_2S + 3/2\ O_2 \leftrightarrow 3S + 3H_2O \qquad (10.5)$$

Figure 10.5 shows a process flow diagram for a typical sulfur recovery unit.

The thermal step is carried out in the burner, which operates under temperatures higher than 900°C. Close to one-third of H_2S is converted in SO_2, following Reaction 10.1, which is endothermic. This step is also responsible to destroy the sour gas contaminants as ammonia and hydrocarbons, the thermal step is responsible for 60–70% of sulfur recovery.

The catalytic step is realized in fixed-bed reactors containing TiO_2 or activated alumina as the catalyst. The catalytic step (Reaction 10.4) is slightly endothermic when compared with the thermal step, so it is carried out at lower temperatures (200–350°C). One of the main catalyst developers of the Claus process is Axens.

The cooling of the process stream between the thermal and catalytic steps is realized in a waste heat boiler, producing steam that is sent to consumers in other processes in the refinery. The process applies multiple reaction stages with the removal of produced sulfur among the stages to shift the chemical equilibrium to the products. In units containing three catalytic stages, it is possible to recover 98% of the sulfur contained in the sour gas that is fed to the unit.

To comply with currently SOx emissions regulations, the sulfur recovery units normally need a sulfur recovery efficiency between 99% and 99,5%. To raise the sulfur recovery efficiency, modern sulfur recovery units rely on tail gas treating units, as presented in Figure 10.6.

The tail gas treating unit receives the off-gas from the sulfur recovery unit and converts the remaining SO_2 and other sulfur compounds in H_2S that are sent back to the sulfur recovery unit, raising the sulfur recovery efficiency.

This process consists of a heating step for the residual gas, which raises the temperature of sulfur condensation. This phenomenon occurs in the reactor and supplies the energy needed for the conversion reactions. In this step, hydrogen production still occurs, which acts as a reduction gas to convert the sulfur compounds to H_2S in the catalytic process step.

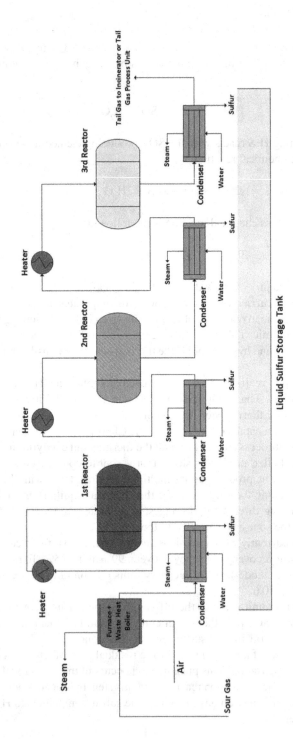

FIGURE 10.5 Process Flow Diagram for a Typical Sulfur Recovery Unit

Environmental Processes

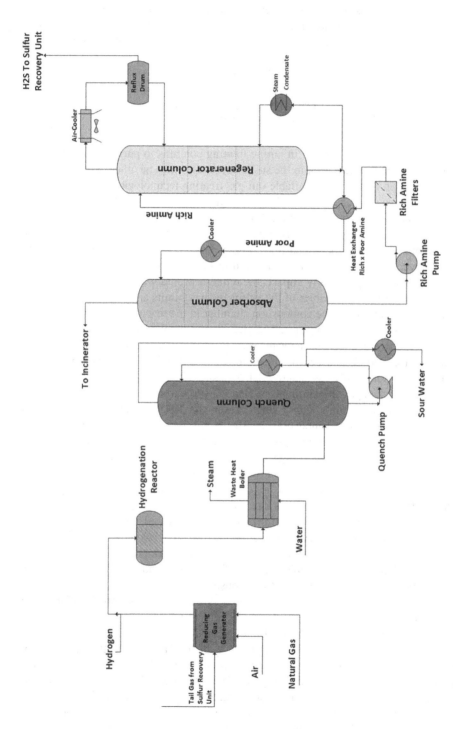

FIGURE 10.6 Process Flow Diagram to Typical Tail Gas Treating Unit

Then the tail gas receives an injection of hydrogen before entering the fixed-bed reactor containing cobalt and molybdenum (Co-Mo) catalysts. In this step, the sulfur compounds are converted to H_2S, according to the following chemical reactions:

$$SO_2 + 3H_2 \rightleftarrows H_2S + 2H_2O$$
$$S + H_2 \rightleftarrows H_2S$$
$$COS + 4H_2 \rightleftarrows CH_4 + H_2S + H_2O$$
$$CS_2 + 4H_2 \rightleftarrows CH_4 + 2H_2S$$

H_2S produced is processed in amine treating columns to purify this compound, and then the H_2S is sent back to the sulfur recovery unit. The other gases are sent to an incinerator. Among the available tail gas treating technologies are SCOT (Shell Off-Gas Treater), licensed by Shell; the Resulf process, developed by Lummus; and the Flexsorb, developed by ExxonMobil.

The sulfur recovery complex is extremely important to adequate modern refinery operation. Normally, operational instabilities and the shutdown of sulfur recovery units force a reduction in the flow rate, and in some cases, the shutdown of refinery processing units significantly impacts the refiner's profitability.

The main process variables of the sulfur recovery units are the air/sour gas ratio and the temperatures of the combustion chamber in the reactors in the catalytic step and the condenser's temperatures. The airflow rate supplied to the process needs to be sufficient to completely burn the hydrocarbons and NH_3 present in the feed stream, plus the necessity to convert the third part of H_2S into SO_2. Combustion chamber temperature is normally sufficiently high to promote the Claus process reactions and to destroy the sour gas contaminants (NH_3 and hydrocarbons). In refineries that apply sour water stripping units with a single tower, the NH_3 content in the sour gas is higher. In this case, the combustion chamber is normally higher (above 1100°C).

One of the main operational problems of sulfur recovery units is related to processing sour gases containing high hydrocarbon content, which raises the air consumption and can lead to soot formation (heavier hydrocarbons), which provokes catalyst deactivation and high pressure drop in the reactors, further requiring higher temperatures in the Claus process's thermal step.

Nowadays, to increasingly minimize the environmental impact of the refining processes, some licensors have devoted their efforts to developing new technologies focused on sulfur and nitrogen recovery from waste gases produced in the refining processes. Among these technologies, one of the most economically attractive is the SNOX technology, developed by Haldor Topsoe Company.

In the SNOX process, sulfur is recovered as a highly concentrated sulfuric acid that can be commercialized directly by the refiner, while nitrogen is eliminated as nitrogen gas (N_2), which is not harmful to the environment.

Sulfur recovery units, as aforementioned, are fundamental to the operation of modern refining complexes. The produced sulfur is normally commercialized to produce sulfuric acid and fertilizers. Beyond minimizing the refining process's environmental impact, sulfur recovery units are capable of adding profitability to the refiner.

Despite the efforts over the years, the crude oil refining industry still presents a great environmental impact, but crude oil derivatives are fundamental to sustaining economic development. In this sense, adequate performance of the environmental units is fundamental to allowing sustainable operation of the refining hardware from economic and environmental points of view. As aforementioned, the unavailability of environmental processing units can lead to the reduction in the processing capacity of the refining hardware aiming to keep under control the atmospheric emissions and leading to great economic losses and the risk of a shortage of crude oil derivatives in the market, in extreme cases. It's always important to consider the relevance of the environmental units to the refining hardware, and the optimization and maintenance priorities of these units need to be put at the same level as processing units.

10.4 WATER AND WASTEWATER TREATMENT TECHNOLOGIES

Despite the efforts to reduce the environmental footprint of its processes, the crude oil processing industry still presents great environmental impacts. One of the main impacts observed in crude oil refining is the high water consumption demanded by the separation and conversion processes applied in the steps to produce derivatives useful to society. It's estimated that on average 250–350 liters of water are consumed per barrel of crude oil processed. However, this consumption depends, among other factors, on the complexity grade of the refinery.

Generally, the water is collected from a water source close to the refinery and, after the adequate treatment, distributed for consumption according to the final application. Normally, the use of water in a refinery is divided into water to fight emergencies, clarified water, industrial water, potable water, and demineralized water. Figure 10.7 presents a simplified diagram of the treatment processes and the water streams produced.

Raw water is collected and sent to a treatment station in the refinery. The portion of the water used to supply the emergency network normally doesn't receive any treatment, being basically composed of raw water. Before the clarification step, the

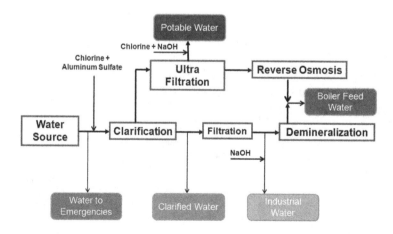

FIGURE 10.7 Simplified Process Flow Diagram of the Water Treatment Processes in Crude Oil Refineries

water receives a dosage of chlorine to promote the oxidation of organic compounds and of aluminum sulfate to favor the coagulation of suspended particles. Clarified water is normally applied to make up the cooling towers of the processing units and as a feed stream to the filtration and ultrafiltration steps. After the filtration steps (normally carried out in sand filters), part of the water stream receives a dosage of caustic soda to control the pH between 7,5 and 8,5. This portion of the water is called industrial water and is used directly in the process like water wash in crude oil desalters or reactors decoking fluid in delayed coking units.

The potable water receives an additional treatment with chlorine for disinfection and caustic soda injection to keep the pH in the range of 7–8,5 and is applied in general use in the administrative and industrial areas of the refinery.

The ultrafiltration process applies membranes with pores in the range of 0,01–0,001 μm capable of promoting the separation of the major part of viruses, bacteria, and high-molecular-weight compounds. After the ultrafiltration step, the filtrated water passes through the reverse osmosis step that applies membranes even more selective and capable of removing particles in the range of 1–10 angstroms. Posteriorly, the water is applied to feed boilers, which will produce steam for the process and the stream of demineralized water.

The demineralization process is composed of beds of activated carbon to remove the chlorine from the water to avoid damage to the ionic exchange resins. Posteriorly, the process uses vessels filled with polymers chemically actives that promote the exchange of the ions between the resin and the water to be treated, as described in Figure 10.8.

The process is based on the substitution of ions from the water stream by equivalent quantities of ions with the same electric charge by the resin. The demineralized water, as well as the permeated water (from the reverse osmosis), can be applied directly as boiler feed water. The quality parameters controlled in the boiler feed water are pH (8,5–9,8), conductivity that is controlled below 1,5 mho, and the silicon content that can't be superior to 0,05 ppm.

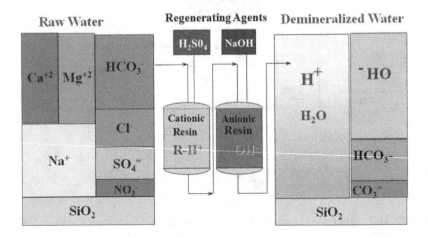

FIGURE 10.8 Demineralized Water Production Process

Environmental Processes

One of the most relevant issues for refiners in recent years has been the reduction of water consumption in processes. The lower water intake for industrial applications, besides reducing the operational costs for the refiners, allows more water to be available for nobler purposes, such as human consumption. This fact is relevant mainly in regions where there is a lack of good-quality water.

One of the most important steps in the crude oil refining process is the treatment of effluents generated during the process of adding value to the streams that will make up the pool of derivatives.

As important to the operational continuity of a refinery as an efficient water treatment is the process of treating wastewater. Environmental norms around the world are increasingly restrictive. These currents have high levels of contaminants, some of them being carcinogenic, like aromatic compounds. Most of the compounds are organic (hydrocarbons, nitrogenated, sulfurous) or ammoniacal and consume a large amount of dissolved oxygen, altering life conditions in these environments. Figure 10.9 presents a process flow diagram for a typical wastewater treatment station.

In addition, the water balance of the refinery will be carried out by the effluent stream, and the total flow will be limited by the treatment capacity of the wastewater system. With this important restriction, any increase in water consumption should be studied considering this premise. On the other hand, current technologies allow the removal of contaminants with such efficiency that the reuse of treated streams is now a reality. The equation of cost of reuse versus cost of raw water should always be considered. However, this ends up being a matter of preservation of a limited resource in practically all regions of the planet.

10.4.1 Oily Sewer

The wastewater streams of a refinery are produced from virtually all processing units. In the distillation unit, the desalting of crude oil precedes the separation process. In this process, water is injected, and oil is subjected to an electric field so that the salts and sediments pass into the aqueous phase (brine) and are sent for treatment. This stream is composed of water, salts, and sediments. The first step of this treatment occurs in a tank with a residence time sufficient to allow separation of the oil phase. The aqueous phase will be treated at the industrial wastewater treatment plant.

Still, in the distillation unit, there is the injection of steam for the rectification. After the cooling of the derivative streams, the steam is condensed and collected as sour water and sent to treatment in a specific processing unit (sour water stripping). The treated stream can be reused in the processing units, with the surplus being sent to the wastewater treatment plant. In the units of catalytic cracking, delayed coking, and hydrotreatment, there is also the generation of sour water that is sent to the same type of treatment.

Cooling towers need to have a limited salt concentration to prevent corrosion or fouling in heat exchangers and piping. This control is done by chemical dosing and purging.

10.4.2 Stormwater Sewer

The streams mentioned previously are continuous and independent of climatic events. In rainy situations, there is also the contribution of the stormwater sewage system of

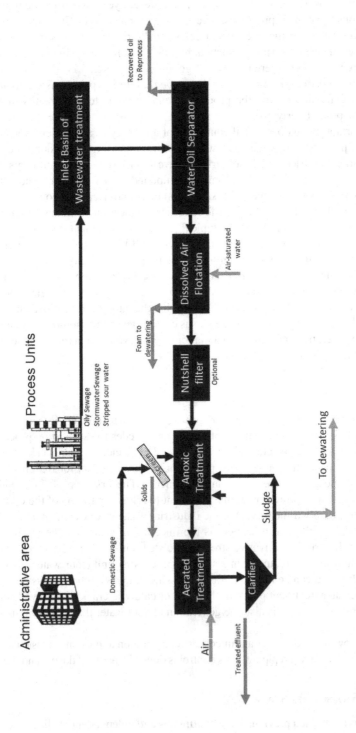

FIGURE 10.9 Wastewater Treatment Station with Activated Sludge Reactor

Environmental Processes

the ground of processing units that are water contaminated with hydrocarbons. For this reason, the treatment plants are designed with a capacity 50% higher than the volume of effluents expected in dry weather.

The stormwater sewer system is an open channel network and covers the whole refinery. Rainwater and stormwater are collected inside the dikes of the storage tanks or in a specific tank and drained to the ETP network.

10.4.3 Domestic Sewage

All the sanitary from toilets and canteens in the refinery (including the administrative building) are connected to this system. This connection is made partly by gravity and partly by pumping.

10.4.4 Steps of Effluent Treatment

The wastewater of oil refining industries is typically biodegradable. The standard scheme of a treatment unit is designed based on physical-chemical processes as primary treatment focusing on free oil and sediment removal, a biological treatment (aerated, anoxic, and/or anaerobic treatment). As a polish treatment, some designs have a microfiltration membrane at the end of the process or dosing an oxidative chemical, like hydrogen peroxide.

The wastewater is routed to the first stage of the treatment of the removal of free oil. The typical concentration of oil and greases can range from 100 mg/L to 1000 mg/L. The equipment responsible for this step is known as a water-oil separator.

10.4.5 API Oil-Water Separator

This is a standard device for separating oil in refineries. The separation is based on the difference in density between the water droplets of suspended oil and the solid particles, creating a three-phase system. The oil is collected on the surface and sent for reprocessing. The solids are collected on the bottom for further dewatering in a device like a centrifugal dewaterer. The aqueous phase, which still contains oily emulsified material, follows for treatment in a flotation process.

10.4.6 Dissolved Air Flotation

Minor oil droplets and very small particle sediments cannot be removed by density difference and require a driving force for efficient removal. A traditional process in oil refineries is dissolved air flotation. In this process, there is a dosage of a chemical to break the emulsion, usually aluminum sulfate, ferric chloride, or another electrolyte capable of inhibiting repulsion between the particles promoting coagulation. With the formation of flakes, it is possible to inject air-saturated water at the bottom of the float. With the reduction of pressure, the air is released, forming small bubbles that drag the flakes to the surface, forming a foam that is removed and sent to the dehydration system. The clarified water is sent to biological treatment. Some designs have a nutshell filtering device focusing on oil removal to avoid foam on the aerated treatment.

10.4.7 Biological Treatment

After oil and sediment removal, biological treatment of wastewater is done under aerobic conditions (excess of free dissolved oxygen) for the reduction of carbon-based organic pollutants and ammonia. The aim is to enhance microbial oxidation of the organic and ammoniacal pollutants using the naturally occurring bacteria in the environment. These bacteria consume a major part of the organic substances as energy and mass source while converting the pollutants into new cells and harmless compounds. In the same way, ammonia is converted to cell proteins and nitrates/nitrites in wastewater bulk.

Inactivated microbes result in sludge that must be removed periodically to maintain the treatment activity.

The basic biochemical reaction for the stabilization of organic and ammoniacal impurities under aerobic conditions by microorganisms in wastewater may be represented as follows:

$$\text{Organic impurities} + \text{Ammonia} + \text{Microbes} + O_2 \rightarrow$$
$$\text{Microbes} + \text{Nitrite} + \text{Nitrate} + CO_2 + H_2O + \text{Waste energy}$$

Nitrite and nitrate are toxic substances and must be removed before discharging. In anoxic conditions, denitrifying bacteria use organic compounds as food but use nitrites and nitrates as oxygen sources. This way, molecular nitrogen is released into the atmosphere as products of the reaction.

$$\text{Organic impurities} + \text{Nitrate} + \text{Nitrite} + \text{Microbes} \rightarrow$$
$$\text{Microbes} + N_2 + CO_2 + H_2O + \text{Waste energy}$$

In aerated systems, bacteria and protozoa are mixed with wastewater and motor-operated aerators or air blowers fed with air. In anoxic systems, aeration is minimum. The bacteria water (mixed liquor) is then sent to the clarifier, where the microbial mass is separated from the water. The bacterial mass returns to the bioreactors to maintain the required level of microbial cell concentration. The rest of the bacteria may be sent to the biological sludge centrifugal dewaterers and disposed after. The water from the clarifier unit goes to the treated water pool, ready for disposal or reuse based on the final quality.

10.4.8 Membrane Bioreactors (MBR)

Membranes are semipermeable barriers that are highly efficient in the separation and purification of components and are amply evident throughout nature. Living cells use the screening properties of membranes to collect and eliminate substances. Since the 1970s, these screening properties have been used in numerous applications of membrane-based technology in separation and purification. A typical treatment station with MBR is presented in Figure 10.10.

Environmental Processes

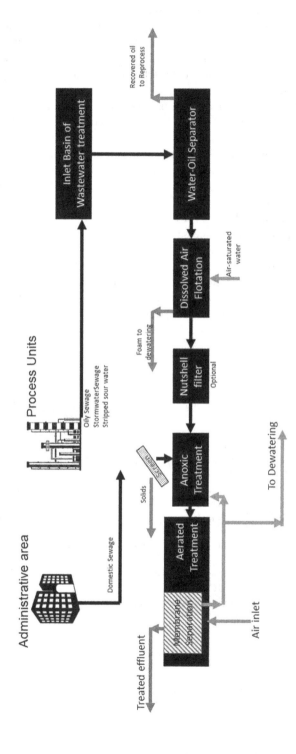

FIGURE 10.10 Wastewater Treatment Station with an MBR Reactor

With the increased availability of membranes of versatile materials, the possibility of integrating membrane separation with conventional treatment nowadays is a reality. This type of membrane-integrated approach in refinery effluent treatment culminates in novel treatment plants with a high degree of process intensification, resulting in safer, more flexible, more compact, and economically viable green processing treatment units.

In this type of plant, the clarifier is substituted by a set of membrane modules, which objective is to separate the sludge from the treated wastewater. This scheme permits a higher quality of final effluent with large possibilities of reusing. The membranes may retain bacteria, viruses, and minor-sized solid particles. Removal of 95% of COD (chemical oxygen demand) and 99% of ammonia can be easily achieved. Typically, the final effluent of a treatment plant of oil refinery wastewater has low COD, BOD (biological oxygen demand), and ammonia. However, the quantities of soluble salts may present high values. To reuse this water in steam generation before a process of demineralization is necessary for desalination, like reverse electrodialysis (RED).

Process sustainability is critical not only to the crude oil processing chain but to any industrial activity. Maintaining the integrity and reliability of these processes is a key factor in ensuring the competitiveness and permanence of refiners in the market in compliance with environmental requirements.

BIBLIOGRAPHY

1. Fahim, M.A., Al-Sahhaf, T.A., Elkilani, A.S. *Fundamentals of Petroleum Refining*. 1st edition, Elsevier Press, 2010.
2. Oliveira, P.C., Silva, M.W. Making the Crude Oil Refining Industry Sustainable – Water and Wastewater Treatment Technologies. 2019, https://www.linkedin.com/pulse/making-crude-oil-refining-industry-sustainable-water-da-silva-mba/?articleId=6543414113140252672.
3. Judd, S., Judd, C. *The MBR Book: Principles and Applications of Membrane Bioreactors for Water and Wastewater Treatment*, Elsevier Press, 2011.
4. Moulijn, J.A., Makkee, M., Van-Diepen, A.E. *Chemical Process Technology*. 2nd edition, John Wiley & Sons Ltd., 2013.
5. Myers, R.A. *Handbook of Petroleum Refining Processes*. 3rd edition, McGraw-Hill, 2004.
6. Riffat, R. *Fundamentals of Wastewater Treatment and Engineering*. 1st edition, CRC Press, 2019.
7. Drinan, J.E., Spellman, J. *Water and Wastewater Treatment: A Guide for a Nonengineering Professional*. 2nd edition, CRC Press, 2012.
8. Cheremisinoff, N. *Environmental Management Systems Handbook for Refineries*. 1st edition, Elsevier Press, 2006.

11 A New Downstream Industry

As presented in previous chapters, the downstream industry is fundamental to ensure added value to the processed crude oil and supply energy to the economic development of nations. Despite its relevance, the downstream industry is facing a transitive period where the pressure to minimize the environmental footprint of the crude oil processing chain, new technologies, and new behavior patterns of society are producing significant changes in the downstream industry.

The COVID-19 pandemic had a great impact on the downstream industry as it caused a drastic reduction in crude oil demand due to the mobility restrictions. Despite the impact of the crisis on all players of the downstream industry, the data from IEA (International Energy Agency) presented in Figure 11.1 indicates that the most integrated players experienced less impact than the players relying on exclusively transportation fuel as a revenue source.

The data presented in Figure 11.1 reinforces the advantage of petrochemical integration in the downstream industry as an edge, considering the falling demand for transportation fuel at a global level.

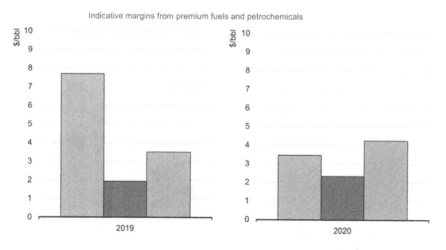

FIGURE 11.1 Impact of the COVID-19 Pandemic on the Refining Margins

Source: IEA (2021)

The improvement in fuel efficiency and the growing market of electric vehicles tend to decrease the participation of transportation fuel in the global crude oil demand. New technologies like additive manufacturing (3D printing) have the potential to produce a great impact on the transportation demands, leading to even more impact on the transportation fuel demand. Furthermore, the higher availability of lighter crude oil favors the oversupply of lighter derivatives that facilitate the production of petrochemicals against transportation fuels, as well as the higher added value of petrochemicals in comparison with fuels. It's interesting to note that even in a pandemic scenario, the sales of electric vehicles grew close to 40% in 2020 and grew even more in 2021, according to data from International Energy Agency (IEA), presented in Figure 11.2.

Facing these challenges, the search for alternatives that ensure the survival and sustainability of the refining industry became constant for refiners and technology developers. Due to their similarities, better integration of refining and petrochemical production processes appears as an attractive alternative. Some forecasts indicate that the chemical market will be doubled in 2030 in comparison with the 2016 level. In general, the refiners focused on transportation fuel are capable of adding close to $15 per processed barrel, while the players focused on petrochemicals can achieve $30 per processed barrel. In this sense, the petrochemical integration allows participation not only in a most profitable market but also in a growing and most resilient market when compared with the transportation fuels.

Despite the advantages, it's important to consider that the integration between refining and petrochemical assets increases the complexity, requires capital spending,

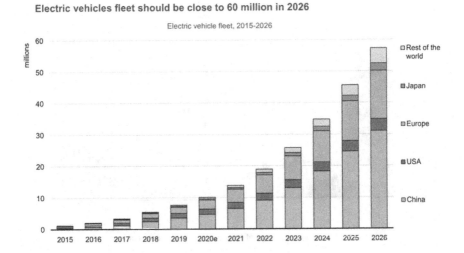

FIGURE 11.2 Evolution of the Electric Car Fleet

Source: IEA (2021)

and affects the interdependency of refineries and petrochemical plants. These facts need to be deeply studied and analyzed case by case.

According to the recent forecasts, it is expected that the Asian market to respond by 90% of the expected growth of world crude oil consumption in the 2019–2026 period, and the most of this growth is related to petrochemical demand. According to data from Asian Downstream Insights (2021), an annual growth of 4,25% is expected in the petrochemical demand in the Asian market in the next five years. This scenario is leading the players of the Asian downstream sector to lead the world's efforts for petrochemical integration. This is translated into high capital spending on crude-to-chemicals refineries, especially in China.

11.1 WHAT IS PETROCHEMICAL INTEGRATION?

The focus of the closer integration between refining and petrochemical industries is to promote and seize the synergies and existing opportunities between both downstream sectors to generate value for the whole crude oil production chain. Table 11.1 presents the main characteristics of the refining and petrochemical industry and the synergies' potential.

As aforementioned, the petrochemical industry has been growing at considerably higher rates when compared with the transportation fuel market in the last few years. Additionally, the industry represents a noble destiny and is less environmentally aggressive than crude oil derivatives. The technological bases of the refining and petrochemical industries are similar, which leads to possibilities of synergies capable of reducing operational costs and adding value to derivatives produced in the refineries.

Figure 11.3 presents a block diagram that shows some integration possibilities between refining processes and the petrochemical industry.

Process streams considered with low added value to refiners like fuel gas (C_2) are attractive raw materials to the petrochemical industry, as well as streams considered

TABLE 11.1
Refining and Petrochemical Industry Characteristics

Refining Industry	Petrochemical Industry
Large feedstock flexibility	Raw material from naphtha/NGL
High capacities	Higher operation margins
Self-sufficient in power/steam	High electricity consumption
High hydrogen consumption	High availability of hydrogen
Streams with low added value (unsaturated gases and C_2)	Streams with low added value (heavy aromatics, pyrolysis gasoline, C_4)
Strict regulations (e.g., benzene in gasoline)	Strict specifications (hard separation processes)
Transportation fuel demand declining at a global level	High-demand products

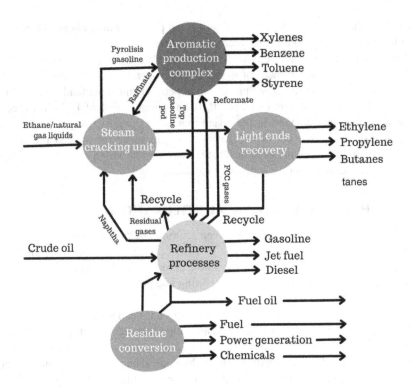

FIGURE 11.3 Synergies between Refining and Petrochemical Processes

residual to petrochemical industries (butanes, pyrolysis gasoline, and heavy aromatics) can be applied to refiners to produce high-quality transportation fuels, this can help the refining industry meet the environmental and quality regulations to derivatives.

The integration potential and the synergy among the processes rely on the refining scheme adopted by the refinery and the consumer market. Processing units such as FCC and catalytic reforming can be optimized to produce petrochemical intermediates to the detriment of streams that will be incorporated into the fuel pool. In the case of FCC, the installation of units dedicated to producing petrochemical intermediates, called petrochemical FCC, to reducing to the minimum the generation of streams to produce transportation fuels. However, the capital investment is high once the severity of the process requires the use of material with the noblest metallurgical characteristics.

IHS Markit proposed a classification of the petrochemical integration grades, as presented in Figure 11.4.

According to the classification proposed, crude-to-chemicals refineries are considered the maximum level of petrochemical integration, where the processed crude oil is totally converted into petrochemicals.

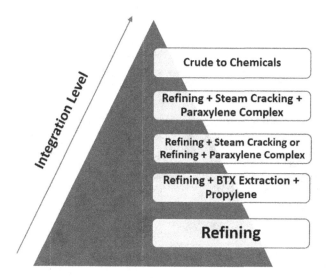

FIGURE 11.4 Petrochemical Integration Levels

Source: Modified from IHS Markit, 2018

11.2 MORE ADDED VALUE TO THE PROCESSED CRUDE: INTEGRATED REFINING SCHEMES

Historically, the refining industry growth was sustained and focused on transportation fuels. This can explain the profile of the traditional refining schemes. Nowadays, the downstream industry is facing a trend of reduction in transportation fuel demand, followed by growing demand for petrochemicals. This fact is the main driving force in promoting the change of focus in the downstream industry.

The growing market for petrochemicals has led some refiners to look for closer integration between refining and petrochemicals assets to reach more adherence with the market demand, improve revenues, and reduce the operation costs. To reach this goal, the refiners are implementing the most integrated refining schemes, as presented in Figure 11.5.

As presented in Figure 11.5, the integrated refining scheme relies on flexible refining technologies such as catalytic reforming and FCC units that are capable of reaching the production of high-quality petrochemicals and transportation fuels, according to the market demand. A more integrated refining configuration allows the maximization of petrochemicals, raising the refining margins and ensuring higher value addition to the processed crude oil. Another fundamental competitive advantage is the operational flexibility reached through the integrated refining configurations, allowing the processing of discounted and cheaper crude oil, raising the refining margins even more. The refining configuration presented in Figure 11.5 is considered a highly integrated refining hardware according to the classification proposed in Figure 11.4, one degree below the crude-to-chemicals refining asset.

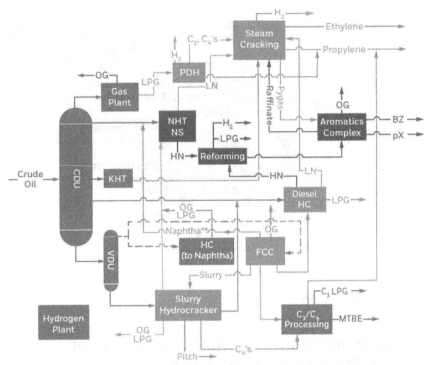

**Light Naphtha (LN) to Steam Cracker, Heavy Naphtha (HN) to Platformer for Max Petrochemicals
Offgas (OG) and LPG are routed to Steam Cracker for Max Petrochemicals

FIGURE 11.5 Example of an Integrated Refining Focusing on Petrochemicals Scheme by UOP (with Permission)

Source: Honeywell UOP (www.uop.honeywell.com)

11.3 CRUDE OIL TO CHEMICALS: ZERO FUEL AND MAXIMUM ADDED VALUE

Due to the increasing market and higher added value and the trend of reduction in transportation fuel demand, some refiners and technology developers have dedicated their efforts to developing crude-to-chemicals refining assets. One of the big players that have invested in this alternative is Saudi Aramco. The concept is based on the direct conversion of crude oil to petrochemical intermediates, as presented in Figure 11.6.

The process presented in Figure 11.6 is based on the quality of crude oil and deep conversion technologies like high-severity or petrochemical FCC units and deep hydrocracking technologies. The processed crude oil is light with low residual carbon, which is a common characteristic in the Middle East crude oil. The processing scheme involves the deep catalytic conversion process to reach maximum conversion to light olefins. In this refining configuration, the petrochemical FCC units have a

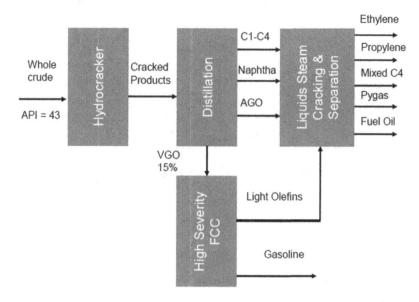

FIGURE 11.6 Saudi Aramco Crude-to-Chemicals Concept

Source: IHS Markit, 2017, with Permission

key role in ensuring high added value to the processed crude oil. An example of FCC technology developed to maximize the production of petrochemical intermediates is the PetroFCC process by UOP. This process combines a petrochemical FCC and separation processes optimized to produce raw materials for the petrochemical process plants, as presented in Figure 11.7. Other available technologies are the HS-FCC process, commercialized by Axens, and the Indmax process, licensed by Lummus.

It's important to consider that both technologies presented in Figure 11.7 are based on petrochemical FCC units that have a special design due to the most severe operating conditions.

For petrochemical FCC units, the reaction temperature reaches 600°C, and a higher catalyst circulation rate raises the gas production, which requires a scaling up of the gas separation section. The higher thermal demand makes the operation of the catalyst regenerator advantageous in total combustion mode, leading to the necessity of installing a catalyst cooler system.

The installation of petrochemical catalytic cracking units requires a deep economic study considering the high capital investment and higher operational costs. However, some forecasts indicate growth of 4% per year in the market of petrochemical intermediates until 2025. This scenario can attract capital investment to raise the market share in the petrochemical sector, allowing the refiner a favorable competitive positioning through the maximization of petrochemical intermediates. Figure 11.8 presents a block diagram showing a case study demonstrating how the petrochemical FCC unit—in this case, the Indmax technology, by Lummus—can maximize the yield of petrochemicals in the refining hardware.

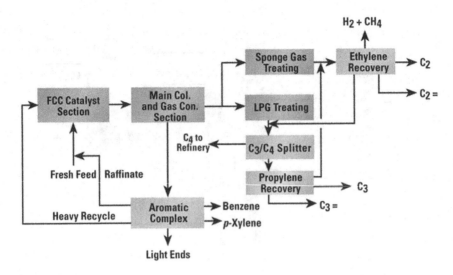

FIGURE 11.7 PetroFCC Process, by UOP (with Permission)
Source: Honeywell UOP (www.uop.honeywell.com)

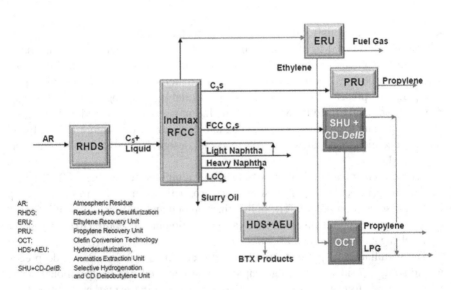

FIGURE 11.8 Olefin Maximization in the Refining Hardware with Indmax FCC Technology, by Chevron Lummus (with Permission)

A New Downstream Industry

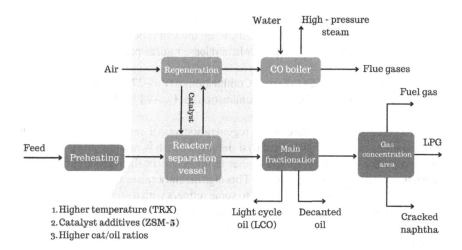

FIGURE 11.9 Optimization of Process Variables in FCC Units to Improve the Yield of Petrochemical Intermediates

In refining hardware with conventional FCC units, it's possible to operate under operating conditions to maximize petrochemicals (ethylene, propylene, C_4s). In this operation mode, the FCC unit operates under high severity, translated to high operation temperature (TRX) and high catalyst/oil ratio. The catalyst formulation considers higher catalyst activity through the addition of ZSM-5 zeolite. There is the possibility of a reduction in the total processing capacity due to the limitations in blowers and cold area capacity.

An improvement is observed in the octane number of cracked naphtha despite a lower yield due to the higher aromatic concentration in the cracked naphtha. In some cases, the refiner can use the recycled cracked naphtha to improve the LPG yield even more.

In the maximum LPG operation mode, the main restrictions are the cold area processing capacity, metallurgic limits in the hot section of the unit, treating section processing capacity, and the top systems of the main fractionating column. In markets with falling demand for transportation fuels, this is the most common FCC operation mode.

By changing the reaction severity, it is possible to maximize the production of petrochemical intermediates, mainly propylene, in conventional FCC units, as shown in Figure 11.9.

The use of FCC catalyst additives, such as ZSM-5, can increase unit propylene production by up to 9%. Despite the higher operating costs, the higher revenues from the higher added value of derivatives should lead to a positive financial result for the refiner, according to current market projections. A relatively common strategy also applied to improve the yield of LPG and propylene in FCC units is the recycling of cracked naphtha, leading to an over-cracking of the gasoline-range molecules.

These operating conditions require the installation of a catalyst cooler system, which raises the processing unit profitability through the total conversion enhancement and selectivity to noblest products such as propylene and naphtha against gases and coke production. The catalyst cooler is necessary when the unit is designed to operate under total combustion mode due to the higher heat release rate, as presented here:

$$C + \tfrac{1}{2} O_2 \rightarrow CO \text{ (Partial Combustion)} \quad \Delta H = -27 \text{ kcal/mol}$$
$$C + O_2 \rightarrow CO_2 \text{ (Total Combustion)} \quad \Delta H = -94 \text{ kcal/mol}$$

In this case, the temperature of the regeneration vessel can reach values close to 760°C, leading to higher risks of catalyst damage, which is minimized by installing a catalyst cooler. Furthermore, the higher temperature in the regenerator requires materials with the noblest metallurgy. This significantly raises the installation costs of these units, which can be prohibitive to some refiners with restricted capital access.

Another key refining technology for crude-to-chemicals refineries is the hydrocracking units. Despite the high performance, the fixed-bed hydrocracking technologies cannot be economically effective in treating crude oil directly due to the possibility of a short operating life cycle. Technologies that use ebulliated bed reactors and continuum catalyst replacement allow higher campaign periods and higher conversion rates. Among these technologies, the most known are the H-Oil and Hyvahl technologies, developed by Axens; the LC-Fining process by Chevron Lummus; and the Hycon process, by Shell Global Solutions. These reactors operate at temperatures above 450°C and pressures until 250 bar. Figure 11.10 presents a typical process flow diagram for an LC-Fining processing unit, developed by Chevron Lummus, while the H-Oil process, by Axens, is presented in Figure 11.11.

Catalysts applied in hydrocracking processes can be amorphous (alumina and silica-alumina) and crystalline (zeolites) and have bifunctional characteristics as the cracking reactions (in the acid sites) and hydrogenation (in the metal sites) occur simultaneously.

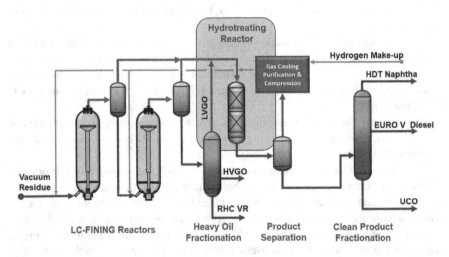

FIGURE 11.10 Process Flow Diagram for the LC-Fining Technology by CLG (with Permission)

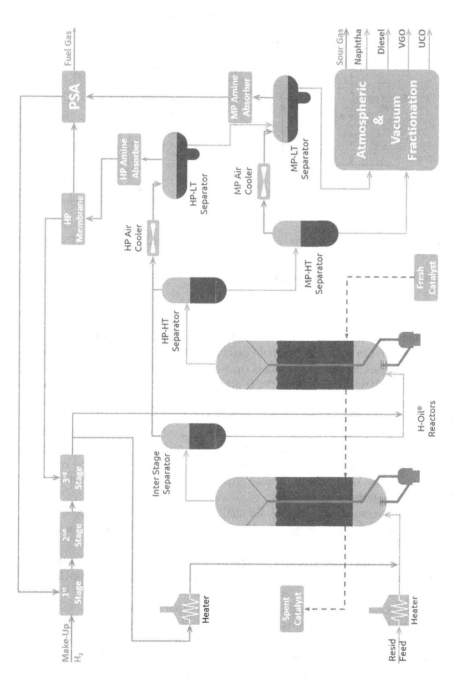

FIGURE 11.11 Process Flow Diagram for H-Oil Process, by Axens (with Permission)

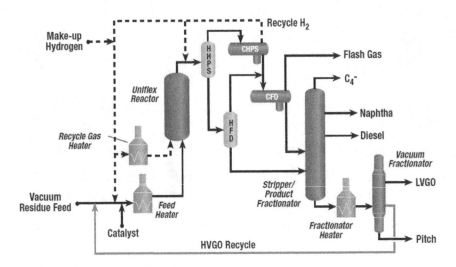

FIGURE 11.12 Process Flow Diagram for Uniflex Slurry Phase Hydrocracking Technology, by UOP (with Permission)

Source: Honeywell UOP (www.uop.honeywell.com)

An improvement in relation to ebulliated bed technologies is the slurry phase reactors, which can achieve conversions higher than 95%. In this case, the main available technologies are the HDH process (hydrocracking-distillation-hydrotreatment), developed by PDVSA-Intevep; the VCC process (Veba Combi-Cracking), commercialized by KBR; the EST process (Eni Slurry Technology), developed by the Italian state oil company Eni; and the Uniflex technology, developed by UOP.

In the slurry phase hydrocracking units, the catalysts in injected with the feedstock and activated in situ while the reactions are carried out in slurry phase reactors, minimizing the reactivation issue and ensuring higher conversions and operating life cycle. Figure 11.12 presents a basic process flow diagram for the Uniflex slurry hydrocracking technology by UOP.

Other commercial technologies for the slurry hydrocracking process are the LC-Slurry technology, developed by Chevron Lummus, and the Microcat-RC process, by ExxonMobil.

For this side, the steam cracking process has a fundamental role in the petrochemical industry. Nowadays, most of the light olefins, light ethylenes, and propylenes are produced through the steam cracking route. Steam cracking consists of a thermal cracking process that can use gas or naphtha to produce olefins.

Naphtha to steam cracking is composed basically of straight-run naphtha from crude oil distillation units. Normally, to meet the requirements as petrochemical naphtha, the stream needs to present high paraffin content (higher than 66%).

Due to their relevance, great technology developers have dedicated their efforts to improving steam cracking technologies over the years, especially related to steam cracking furnaces. Companies like Stone & Webster, Lummus, KBR, Linde, and Technip develop technologies for the steam cracking process. One of the most known steam cracking technologies is the SRT process (Short Residence Time), developed by Lummus, which uses a reduced residence time to minimize the coking process and ensure a higher operational life cycle. Another commercial technology dedicated to optimizing the yield of ethylene is the Score technology, developed by KBR and ExxonMobil, which combines a selective steam cracking furnace with a high-performance olefin recovery section.

Cracking reactions occur in the furnace tubes. The main concern and limitation to the operating life cycle of steam cracking units is coke formation in the furnace tubes. The reactions are carried out under high temperatures, between 500°C to 700°C, according to the characteristics of the feed. For heavier feeds like gas oil, a lower temperature is applied to minimize coke formation. The combination of high temperatures and low residence time are the main characteristic of the steam cracking process. Despite the possibility of operating with naphtha, nowadays the steam cracking operators have chosen to operate with ethane or LPG against naphtha due to the competitive prices related to the new sources of NGL (Natural Gas Liquid). Despite this trend over the last few years, in markets where gasoline surplus is observed, naphtha can still be an attractive alternative as feedstock to steam crackers.

According to some forecasts, the demand for propylene will increase from 130 million metric tons in 2020 to around 190 million metric tons in 2030. Facing the increasingly light feed to refineries and steam cracking units, which tend to favor the ethylene production to the detriment of propylene, propylene demand tends to be supplied by on-purpose propylene production routes like propane dehydrogenation, methanol-to-olefin (MTO) processes, and olefin metathesis.

As quoted previously, some technology developers are dedicating their efforts to developing commercial crude-to-chemicals refineries. Figure 11.13 presents the concept of the crude-to-chemicals refining scheme by Chevron Lummus.

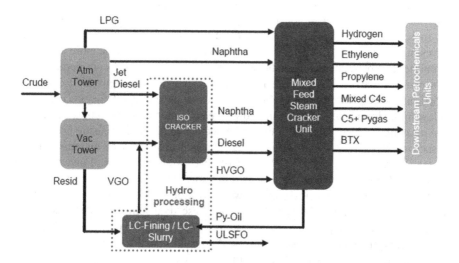

FIGURE 11.13 Crude-to-Chemicals Concept by Chevron Lummus Company (with Permission)

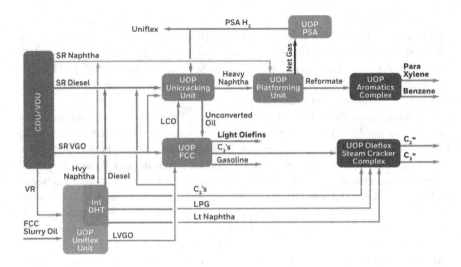

FIGURE 11.14 Integrated Refining Configuration Based on the Crude-to-Chemicals Concept by UOP (with Permission)

Source: Honeywell UOP (www.uop.honeywell.com)

Other great refining technology developers like UOP, Shell Global Solutions, ExxonMobil, Axens, and others are developing crude-to-chemicals technologies, reinforcing that this is a trend in the downstream market. Figure 11.14 presents a highly integrated refining configuration capable of converting crude oil to petrochemicals, developed by UOP.

As presented in Figure 11.14, the production focus change to the maximum adding value to crude oil through the production of high-added-value petrochemical intermediates or chemicals for general purpose, leading to a minimum production of fuels. As aforementioned, big players such as Saudi Aramco have made great investments in crude-to-chemicals technologies to achieve even more integrated refineries and petrochemical plants, raising their competitiveness considerably in the downstream market. The major technology licensors such as Axens, UOP, Lummus, Shell, and ExxonMobil have applied resources to develop technologies capable of allowing a closer integration in the downstream sector to allow refiners to extract the maximum added value from the processed crude oil, an increasing necessity in a scenario where the refining margins are under pressure. Based on data from the Catalyst Group Company (TCGR), in 2019, there were some capital investments in crude-to-chemicals projects, as presented in Table 11.2.

Some of these capital investments were postponed due to the economic crisis provoked by the COVID-19 pandemic, but these data reinforce the trend in the market. It's interesting to note that close to 64% of the global crude-to-chemicals investments are made by Asian players. A typical concern related to the crude-to-chemicals enterprises is related to the operation costs in comparison with the traditional routes. Figure 11.15 presents a comparative study of the operation costs of

TABLE 11.2
Crude-to-Chemicals Investments

Company	Location	Capital Spending (USD Billion)	Enterprise Type
Zhejiang Petroleum and Chemical	China	26	Greenfield
Hengli Petrochemical	China	11	Greenfield
Shengong Petrochemical	China	11.84	Greenfield
Ningbo Zhongjin Petrochemical	China	5.0	Revamp
Huajin Aramco Petrochemical	China	10+	Greenfield
SABIC/Fuhaichuang Petrochemical	China	NA	Greenfield
Sinopec/SABIC (Tianjin Petrochemical	China	45	Revamp
Petrochina	China	-	Revamp
Petrochina	China	-	Revamp
CNOOC	China	-	Revamp
Sinopec	China	2.8	Greenfield
Sinopec	China	4.2	Greenfield
Sinopec	China	4.26	Greenfield
Total Capital Investments in China		**120.1**	
Other Asia			
Hengyi Group	Brunei	20	Greenfield
Saudi Aramco/ADNOC/India Consortium	India	44	Greenfield
Petronas/Saudi Aramco (RAPID)	Malaysia	2.7	Greenfield
ExxonMobil (Singapore Chemical Plant)	Singapore	< 1	Revamp
Pertamina/Rosneft	Indonesia	15	Greenfield
Total Capital Investments in Other Asia		**82.7**	
Middle East			
ADNOC	UAE	45	Revamp
Saudi Aramco/SABIC	Saudi Arabia	30	Greenfield
Saudi Aramco/Total	Saudi Arabia	5	Greenfield
KNPC/KIPIC (Al-Zour Refinery)	Kuwait	13	Greenfield
Oman Oil Company/Kuwait Petroleum International (Duqm Refinery)	Oman	15	Greenfield
Total Capital Investments in Middle East		**108**	
Europe			
MOL Group	Hungary	4,5	Revamp
Total Capital Investments		**315**	

Source: Based on the Catalyst Group, 2019, with Permission

the Hengli crude-to-chemicals enterprise in relation to traditional ethylene production routes.

It's important to consider that the cost composition evolves several factors, and this scenario can be different according to the local business environment. Figure 11.16 presents a comparison between the petrochemical yields of traditional refineries, a benchmark integrated refinery, and the Hengli crude-to-chemicals complex, according to data from IHS Markit.

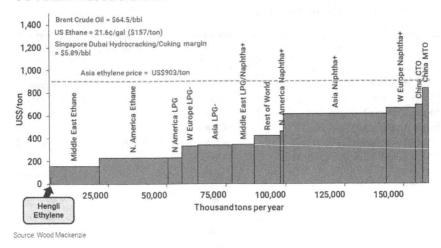

FIGURE 11.15 Ethylene Production Cost Comparison

Source: Wood Mackenzie, 2019, with Permission

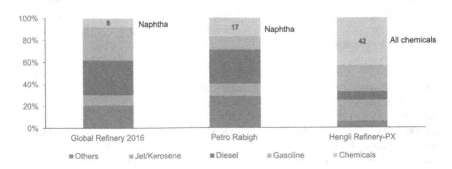

FIGURE 11.16 Petrochemical Yield Comparison

Source: IHS Markit, 2018, with Permission

Analyzing Figure 11.16, it's possible to note the higher added value reached in crude-to-chemicals refineries when compared even with highly integrated refineries. According to data from Wood Mackenzie Company presented in Figure 11.17, the highly integrated refiners can rise from $0,68 to $2,02/bbl. Still, according to Wood Mackenzie, the Asian market presents the major concentration of integrated refining plants.

As aforementioned, faced with the current trend of reduction in transportation fuel demand at the global level, the capacity of maximum adding value to crude oil can be a competitive differential to refiners. Due to the high capital investment

A New Downstream Industry

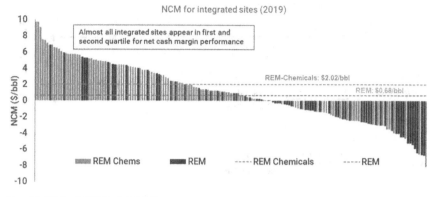

FIGURE 11.17 Average Margins of Integrated Refining Sites
Source: Wood Mackenzie, 2021, with Permission

needed for the implementation that allows the conventional refinery to achieve the maximization of chemicals, capital efficiency also becomes an extremely important factor in the current competitive scenario and the operational flexibility related to the processed crude oil slate.

11.3.1 Available Crude-to-Chemicals Routes

Nowadays, there are three technically available routes that are being considered for capital investments to crude-to-chemicals refining complexes. Figure 11.18 presents the concepts based on the information from IHS Markit.

The conventional routes consider the processing of crude oil in a conventional crude oil refinery, producing petrochemical intermediates like naphtha, which is supplied to a petrochemical asset like a steam cracking unit. The ExxonMobil route is based on the direct feed of selected crude oil, normally light and low-contaminant crudes, to petrochemical assets, while the Chinese enterprise Hengli Zhejiang Shenghong Henyi considers the feed of mixed crude oil slate to a crude-to-PX (paraxylene) complex in order to secure the domestic Chinese market, which presents high demand for light aromatics (BTX). A conventional highly integrated refining hardware is capable of achieving 15–20% of petrochemical yield, while a crude-to-chemicals refinery can reach up to 40%, as presented in Figure 11.16.

As aforementioned, the Aramco/Sabic concept is based on high-complexity refining hardware to convert selected crude oil (light) to maximize the yield of petrochemical intermediates, mainly light olefins.

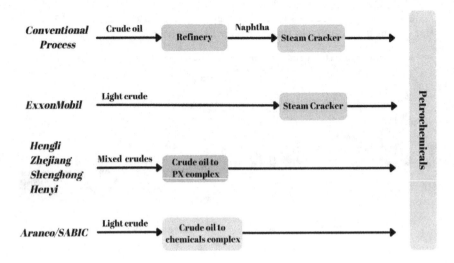

FIGURE 11.18 Crude-to-Chemicals Refining Concepts

Source: Modified from IHS Markit, 2019

Despite the advantages presented by closer integration between refining and petrochemical assets, it's important to understand that the players of the downstream industry are facing a transitive period where, as presented in Figure 11.1, transportation fuel is responsible for a great part of the revenues. In this business scenario, it's necessary to define a transition strategy where the economic sustainability achieved by the current status (transportation fuels) needs to be invested in building the future (maximize petrochemicals). Keeping an eye out only on the future or only on the present can be a competitive mistake.

11.3.2 The Residue Upgrading Technologies in the Integration of Refining and Petrochemical Assets

The hard scenario faced by the whole oil and gas industry requires alternative actions to ensure maximum added value to crude oil. The downstream sector faces lower demand for crude oil derivatives and is put under pressure by the refining margins and competitiveness, mainly from refiners relying on refining hardware with low complexity and poor capacity of bottom barrel conversion and capable of producing lower yields of added-value derivatives. In the current scenario, the players of the downstream sector can consider processing heavier crude slates to improve their refining margins as these crudes present low cost when compared with lighter crude oil, or in some cases, there is a great availability of heavy crudes.

In this scenario, processing units capable of adding value to bottom barrel streams, improving the quality of crude oil residue streams (vacuum residue, gas oils, etc.), or converting them to higher-added-value products gain strategic importance, mainly in countries that have large heavy crude oil reserves, like Venezuela, Canada, and Mexico. These processing units are fundamental to comply with the environmental and quality regulations, as well as to ensure the profitability and competitiveness of refiners by raising the refining margin. The necessity to add value to bottom barrel streams increased even more after January 2020, when the IMO 2020 started, which requires a deep reduction in the sulfur content of the marine fuel oil (bunker) from 3,5% (m.m) to 0,5% (m.m).

Available technologies for processing bottom barrel streams involve processes that aim to raise the H/C relation in the molecule, either through reducing the carbon quantity (processes based on carbon rejection) or through hydrogen addition. Technologies that involve hydrogen addition encompass hydrotreating and hydrocracking processes, while technologies based on carbon rejection refer to thermal cracking processes like visbreaking, delayed coking and fluid coking, catalytic cracking processes like FCC, and physical separation processes like solvent deasphalting units.

Another fact that raises the relevance of the residue upgrading technologies is the growing demand for petrochemicals. This fact requires a higher bottom barrel conversion capacity to convert residual streams to petrochemical intermediates.

As aforementioned, the residue upgrading units are capable of improving the quality of bottom barrel streams. The main advantage of the integration between residue upgrading and petrochemical units like steam cracking is the higher availability of feeds with better crackability characteristics.

Bottom barrel streams tend to concentrate aromatic and polyaromatic compounds that present uneconomically performance in steam cracking units due to the high yield of fuel oil that presents low added value. Furthermore, aromatics tend to experience a condensation reaction in the steam cracking furnaces, leading to high rates of coke deposition that reduces the operation life cycle and raises the operating costs. In this case, deep conversion units like hydrocracking can offer higher operational flexibility.

Once cracking potential is better for paraffinic molecules, and the hydrocracking technologies can improve the H/C in the molecules converting low-added-value bottom streams like vacuum gas oil to high-quality naphtha, kerosene, and diesel, the synergy between hydrocracking and steam cracking units, for example, can improve the yield of petrochemical intermediates in the refining hardware. An example of a highly integrated refining configuration relying on hydrocracking is presented in Figure 11.5.

Taking into account the recent trend of reduction in transportation fuel demand, followed by the growth of the petrochemical market, the presence of hydrocracking units in the refining hardware raises the availability of high-quality intermediate streams capable of being converted into petrochemicals, an attractive way to maximize the value addition to processed crude oil in the refining hardware. As presented in Figure 11.14, the synergy between carbon rejection and hydrogen addition technologies like FCC and hydrocracking units can offer an attractive alternative. Sometimes the hydrocracking and FCC technologies are faced by competitors' technologies in the refining hardware due to the similarities of feed streams that are processed in these units. In some refining schemes, the mild hydrocracking units can be applied as a pretreatment step to FCC units, especially to bottom barrel streams with a high metal content that are severe poisons to FCC catalysts. Furthermore, the mild hydrocracking process can reduce the residual carbon to FCC feed, raising the performance of the FCC unit and improving the yield of light products, like naphtha, LPG, and olefins.

Considering the great flexibility of deep hydrocracking technologies that are capable of converting feed streams varying from gas oils to residues, an attractive alternative to improve the bottom barrel conversion capacity is to process in the hydrocracking units the uncracked residue in the FCC unit to improve the yield of high-added-value derivatives in the refining hardware, mainly middle distillates like diesel and kerosene.

Being capable of adding value to bottom barrel streams can be a great competitive advantage among the refiners, especially considering the possibility of processing heavier crude oil, which presents lower costs and offers the opportunity to raise the refining margins.

Beyond this, the relevance of residue upgrading technologies raised, even more, after the start of the new regulation on bunker (marine fuel oil), the IMO 2020. Once the low-sulfur crude oil becomes scarce, the refiners need to looking for alternative routes to add value to their crude oil reserves and supply the new marine fuel oil, ensuring participation in a profitable market.

As aforementioned, the capacity to add value to the bottom barrel streams is a competitive differential in the refining industry, and this differential tends to be even more relevant in the market scenario with the IMO 2020, especially in markets with easy access to high-sulfur crude oil. The scenario faced by the players of the downstream industry requires even more competitive capacity to ensure higher value addition to the processed crude oil, mainly considering the current trend of reduction in transportation fuel demand followed by the growing market for petrochemicals that requires a higher conversion capacity in the refining hardware to ensure higher yields of added-value derivatives. In this scenario, highly integrated refining configurations based on residue upgrading and flexible refining technologies can be economically attractive. Despite the high capital investment, hydrocracking units and their synergy with other residue upgrading technologies can improve the offer of high-quality intermediates to the petrochemical industry, allowing higher yields of light olefins in the refining hardware and closer integration with petrochemical assets, which is a relevant competitive advantage in the current and short-term scenario of the downstream industry.

11.3.3 Closing the Sustainability Cycle: Plastic Recycling Technologies

As described previously, we are facing a continuous growth of petrochemical demand, and a great part of these crude oil derivatives have been applied to produce common-use plastics. Despite the higher added value and significant economic advantages in comparison with transportation fuels, the main side effect of the growth of plastic consumption is the growth of plastic waste.

Despite the efforts related to the mechanical recycling of plastics, the increasing volumes of plastics waste demand the most effective recycling routes to ensure the sustainability of the petrochemical industry through the regeneration of the raw material. In this sense, some technology developers have dedicated investments and efforts to develop competitive and efficient chemical recycling technologies for plastics.

One of the most applied technologies for plastic recycling is catalytic pyrolysis, where the long-chain polymers are converted into smaller hydrocarbon molecules, which can be fed to steam cracking units to reach a real circular petrochemical industry. Another route is the thermal pyrolysis of plastics; an example of this is the Rewind Mix technology, developed by Axens.

Another promising chemical recycling route for plastics is the hydrocracking of plastic waste. In this case, the chemical principle involves the cracking of carbon-carbon bonds of the polymer under high hydrogen pressure, which lead to the production of stable low-boiling-point hydrocarbons. The hydrocracking route presents some advantages in comparison with thermal or catalytic pyrolysis, as the amount of aromatics or unsaturated molecules is lower than those achieved in the pyrolysis processes, leading to a most stable feedstock to steam cracking or another downstream process, and is more selective, producing gasoline-range hydrocarbons, which can be easily applied in the highly integrated refining hardware.

The chemical recycling of plastics is a great opportunity for technology developers and scientists, especially related to the development of effective catalysts to promote depolymerization reactions which can ensure the recovery of high-added-value molecules like BTX. More than that, the chemical recycling of plastics is an urgent necessity to close the sustainability cycle of an essential industry to our society.

Nowadays, it is still difficult to imagine the global energetic matrix free of fossil transportation fuels, especially in developing economies. Despite this fact, recent forecasts and growing demand for petrochemicals, as well as the pressure to minimize the environmental impact produced by fossil fuels, create a positive scenario and act as the main driving force to the closer integration between refining and petrochemical assets. In extreme scenarios, zero-fuel refineries tend to grow in the middle term, especially in developed economies.

The synergy between refining and petrochemical processes raises the availability of raw material to petrochemical plants and makes the supply of energy to these processes more reliable while ensuring better refining margin to refiners due to the high added value of petrochemical intermediates when compared with transportation fuels. The development of crude-to-chemicals technologies reinforces the necessity of closer integration of refining and petrochemical assets by the brownfield refineries aiming to face the new market that tends to be focused on petrochemicals instead of transportation fuels. It's important to note the competitive advantage of the refiners from the Middle East that have easy access to light crude oil, which can be easily applied in crude-to-chemicals refineries. As presented previously, crude-to-chemicals refineries are based on deep conversion processes that require high capital spending. This fact can put the refiners under pressure with restricted access to capital, again reinforcing the necessity to look for close integration with the petrochemical sector to achieve competitiveness.

On the extreme side of the petrochemical integration trend, there are zero-fuel refineries. As quoted previously, it's still difficult to imagine the downstream market without transportation fuels, but it seems a serious trend, and the players of the downstream sector need to consider a focus change in their strategic plans, like opportunity and threat, mainly considering the pressure on transportation fuel due to the decarbonization necessity and new technologies.

Despite the benefits of petrochemical integration, it's fundamental to consider the necessity to reach a circular economy in the downstream industry. To achieve this goal, the chemical recycling of plastics is essential. As presented previously, there are promising technologies that can ensure the closing of the sustainability cycle of the petrochemical industry.

11.4 RENEWABLES COPROCESSING IN CRUDE OIL REFINERIES

The increasing necessity to reduce the environmental impact produced by fossil fuels have been created a trend of decarbonization of the energetic matrix at a global level, creating then a new challenge to the crude oil production and processing chain.

In this scenario, one of the available alternatives is raising the renewable fuel participation in the energetic matrix and the higher use of renewable raw materials in the feed stream of crude oil refineries, and this fact has led some refining technology licensors to dedicate efforts to develop processes for this purpose.

The adoption of synergies between fossil fuels and renewables in the downstream industry depends on the market where the refiner is inserted, mainly related to the availability of renewable raw materials and the capacity of the installed refining hardware to process the renewable streams.

Despite these limitations, it's important to understand that the renewables are already a reality in the market, contributing to the reduction of the demand for fossil raw material. According to data from International Energy Agency (IEA), the COVID-19 pandemic caused the first contraction in the biofuel market in two decades, as presented in Figure 11.19.

Despite this contraction in 2020, the stricter regulations and policy pressure tend to drive a fast recovery and expansion in the biofuel demand still, according to data from IEA presented in Figure 11.20.

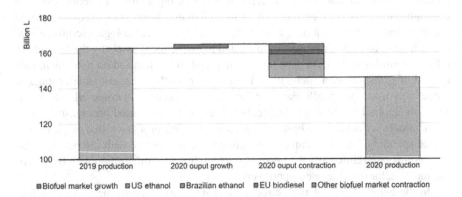

FIGURE 11.19 Biofuel Production in 2019 and the Contraction in 2020

Source: IEA (2020)

A New Downstream Industry

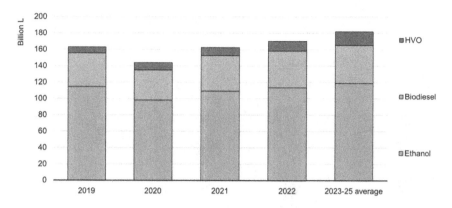

FIGURE 11.20 Global Biofuel Production Forecast
Source: IEA (2020)

Considering these trends, it's possible to estimate the impact of biofuels in the crude oil refining industry, and the coprocessing of renewable raw materials in the traditional crude oil refineries can be an attractive decarbonization strategy. After the COVID-19 pandemic, some refiners decided to convert some refining assets to process renewable raw materials, reinforcing this trend in the new scenario of the downstream industry.

11.4.1 Biofuel Production in Brazil

Brazil has a long tradition in biofuel production. In 1975 due to the petroleum crisis, Brazilian authorities launched an alternative fuel program called PROALCOOL, where the main intention was to support the development of ethanol from sugarcane as automobile fuel in substitution of gasoline to reduce the external dependence of the Brazilian energetic matrix.

According to the Brazilian Petroleum Agency (ANP), in 2019, Brazilian ethanol production reached 35,3 million m^3, considering the volumes of anhydrous and hydrated ethanol. This production reveals consistent growth in the production over the years. Figure 11.21 shows the ethanol production profile over the last few years in the Brazilian market.

Based on data from Figure 11.21, the Brazilian ethanol production growth has an average annual rate of 2,3%. Considering only anhydrous ethanol, the annual growth is even more expressive, reaching 2,6% in the 2010–2019 period. By law, gasoline commercialized in Brazil is 27% anhydrous ethanol in volume, which is applied to improve the gasoline quality (octane boosting) and to ensure participation of renewable fuels in the Brazilian energetic matrix. The hydrated ethanol is commercialized in gas stations as pure fuel for automobiles. Still, according to data from ANP, in 2019 Brazilian production of hydrated ethanol reached 24,9 million m^3, with an average annual growth of 2,1% in the 2010–2019 period.

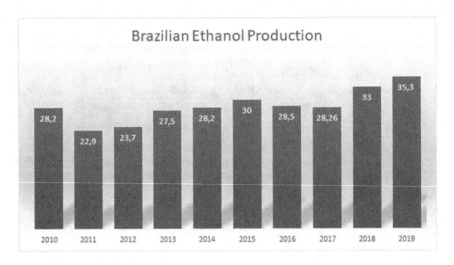

FIGURE 11.21 Evolution of Brazilian Ethanol Production

Source: ANP (2020)

Brazil is a great transportation fuel consumer, and the main driver of the Brazilian economy is the diesel due to the country's dimensions and the transport infrastructure, which relies on road transport. The total production of diesel in the Brazilian market reached close to 41 million m^3 in 2019. By law, the diesel commercialized in the Brazilian territory needs to be 12% biodiesel in volume, and the intention of the Brazilian government is to raise this to 15% in 2023. Figure 11.22 presents the evolution of Brazilian biodiesel production between 2010 and 2019 in m^3 (million).

The main raw material used to produce biodiesel in Brazil is soybean oil, with close to 68% of the total production, followed by animal fat with 11%.

As described earlier, biofuels are fundamental to sustaining the energetic matrix and economic development of Brazil. The blending of anhydrous ethanol with gasoline and biodiesel with diesel represents a kind of strategy to produce cleaner fuels, but this is not the only strategy that is being applied to the refiners aiming to reduce the environmental footprint of the transportation fuels.

An important trend of the energy transition in the downstream industry is the coprocessing of renewable raw materials in crude oil refineries. This strategy involves feeding the renewable raw material directly to the refining process, which represents a more challenging decarbonization strategy.

11.4.2 Challenges of Renewables Coprocessing in Crude Oil Refineries

The use of renewable raw materials in crude oil refineries has been discussed in the last decades. The adoption of synergies between fossil fuels and renewables in the downstream industry depends on the market where the refiner is inserted, mainly related to the availability of renewable raw materials and the capacity of the installed refining hardware to process the renewable streams.

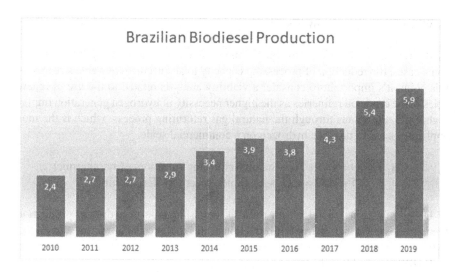

FIGURE 11.22 Evolution of Brazilian Biodiesel Production
Source: ANP (2020)

One of the most common processing routes is the utilization of vegetable or animal oils in the feedstock of conversion or treating units to produce high-quality fuels and petrochemicals. The renewable raw materials can be directly processed together with fossil streams in conversion units like FCC to produce transportation fuel and olefins.

The use of renewable streams also can be applied as a feed stream of hydrotreating units, aiming to produce high-quality fuels like diesel and jet fuel.

Some refiners and technology licensors have developed process technologies that make the higher synergy of renewables possible with the conventional refining industry.

In the petrochemical sector, the production of petrochemical intermediates also has adopted renewables processing routes, such as using ethanol to produce ethylene. Some companies have applied ethylene production through ethanol dehydration since 2010, and some technology licensors have developed processing routes also dedicated to producing ethylene from ethanol.

Despite the advantages of environmental footprint reduction in the refining industry operations, renewables processing presents some technological challenges to refiners.

The renewable streams have a great number of unsaturations and oxygen in their molecules which lead to high heat release rates and high hydrogen consumption. This fact leads to the necessity for a higher capacity of heat removal from hydrotreating reactors to avoid damage to the catalysts. The main chemical reactions associated with the renewable streams hydrotreating process can be represented as follows:

$$R\text{-}CH=CH_2 + H_2 \rightarrow R\text{-}CH_2\text{-}CH_3 \text{ (Olefin Saturation)}$$
$$R\text{-}OH + H_2 \rightarrow R\text{-}H + H_2O \text{ (Hydrodeoxigenation)}$$

Where R represents a hydrocarbon

These characteristics lead to the necessity of higher hydrogen production capacity by the refiners and more robust quenching systems of hydrotreating reactors or, in some cases, the reduction of processing capacity to absorb the renewable streams. At this point, it's important to consider a viability analysis related to the use of renewables in the crude oil refineries as the higher necessity of hydrogen generation implies higher CO_2 emissions through the natural gas reforming process, which is the most applied process to produce hydrogen on a commercial scale.

$$CH_4 + H_2O = CO + 3H_2 \text{ (Steam Reforming Reaction—Endothermic)}$$
$$CO + H_2O = CO_2 + H_2 \text{ (Shift Reaction—Exothermic)}$$

Despite the concern related to the CO_2 emissions due to hydrogen production, there are some cleaner hydrogen production routes that present attractive alternatives to the downstream players like the steam reforming of biomethane, reverse water gas shift (RWGS), and the electrolysis process.

These characteristics lead to the necessity of higher hydrogen production capacity by the refiners and more robust quenching systems of hydrotreating reactors or, in some cases, the reduction of processing capacity to absorb the renewable streams.

This fact leads some technology licensors to dedicate their efforts to looking for alternative routes for hydrogen production on a large scale in a more sustainable manner. Some alternatives pointed out can offer promising advantages:

- Natural gas steam reforming with carbon capture: The carbon capture technology and cost can be a limiting factor among refiners.
- Natural gas steam reforming applying biogas: The main difficulty in this alternative is a reliable source of biogas and its cost.
- Reverse water gas shift reaction ($CO_2 = H_2 + CO$): This is one of the most attractive technologies, mainly to produce renewable syngas.
- Electrolysis: The technology is one of the more promising for the near future.

As aforementioned, hydrogen is a key enabler to the future of the downstream industry, and the development of renewable sources of hydrogen is fundamental to the success of the efforts to the energy transition to a lower carbon profile. In the current scenario, the best alternative for refiners is to optimize the hydrogen consumption, minimizing the operating costs and CO_2 emissions.

Another challenge associated with renewables processing is the cold start characteristics of the derivatives, mainly diesel and jet fuel. The renewable feed streams produce highly paraffinic derivatives after the hydrotreating step. In this sense, the final derivative tends to show a higher cloud point, which can be a severe restriction in colder markets, such as the Northern Hemisphere.

In these markets, refiners tend to apply catalytic beds containing dewaxing catalysts (ZSM-5) in their hydrotreating units or cloud point depressor additives, which can raise the operation costs.

11.4.3 The Hydrotreated Vegetable Oil (HVO): An Attractive Route to Reach "Green Diesel"

As presented earlier, the necessity to build a continuous supply of more sustainable transportation fuel is leading the refiners to consider processing renewable raw materials in the refining hardware to achieve cleaner and fewer carbon fuels. One of the most promising initiatives in this sense is the production of hydrotreated vegetable oil (HVO) to compose the diesel pool of some refineries. The process consists of processing renewable materials like palm oil in conventional diesel hydrotreating units to produce what is called green diesel. At this point, it's interesting to make a differentiation between biodiesel and HVO. Biodiesel is produced through transesterification, producing a mixture of fatty acids and methyl esters. HVO is basically composed of normal paraffin, which is a result of hydrotreating reactions. The great advantage of HVO in comparison with the biodiesel is the similarity of properties in relation to the fossil diesel, the density of HVO tends to be lower than the fossil diesel, and cetane number tends to be high, being a perfect additive in a final mixture. On the other side, the high concentration of normal paraffin leads to worse cold flow characteristics, which can be bypassed through the use of dewaxing beds in hydrotreating reactors, applying ZSM-5 catalysts to control the dimension of the paraffin chain. Due to these characteristics, HVO can be a better blending agent to the final diesel than the traditional biodiesel produced by transesterification.

One of the most relevant challenges of HVO production is the cost of raw material and the choice of this raw material. Another great challenge to HVO production in traditional crude oil refineries is the catalyst applied in the process. Normally, hydrotreating catalysts are composed of metal sulfides like Ni-Mo or Co-Mo carried over by alumina, but the low sulfur concentration and the water production during the hydrotreating reaction of renewable raw materials tend to deactivate these catalysts. An alternative, in this case, is to feed H_2S with the feed stream, but there is always the risk of contaminating the final derivative with high sulfur content. The use of noble metals like Ru and Pt as active metals can solve this problem, but the operating cos can be prohibitive.

Another challenge related to HVO production is the higher heat release in the hydrotreating reactors, which requires well-dimensioned quenching systems. It's important to remember that the conventional hydroprocessing reactors are designed to deal with low contaminant concentrations, while renewable raw materials present a high quantity of unsaturated molecules and oxygen, leading to a high heat release rate.

Another issue related to the coprocessing of renewable raw materials in crude oil refineries is the tendency of water retention in the final derivatives. Due to the chemical structure, biodiesel, for example, tends to retain more than eight times more moisture than fossil diesel, which can lead to issues like microbiological degradation of the fuel in transport and storage systems. The soluble water content in pure biodiesel can reach close to 1800 ppm while the value of the diesel with 20% of biodiesel can reach close to 280 ppm of soluble water, and the diesel with 5% of biodiesel can present up to 150 ppm. This fact will lead the refiners to allow their hardware to remove water from the final derivatives by using draining systems or salt filters to control the moisture content in the final derivatives.

From the point of view of crude oil producers, the renewables coprocessing can be faced as a demand destruction. This threat can be overcome by enjoying the change in the profile of crude oil consumption, where a growing demand for petrochemical intermediates is observed, like ethylene, propylene, and BTX, while the demand for transportation fuel, like gasoline and diesel, present falls.

The energy transition is not a question of choice for the players of the downstream industry; it's a demand from society and a survival question in the middle term. Decarbonization of the energetic matrix requires even more flexibility and agility by refiners to keep and improve their refining margins in the scenario of reduction in the transportation fuel demand and growing demand for petrochemicals. However, as aforementioned, there are available processing technologies capable of allowing the co-processing of renewables and fossil feed streams in crude oil refineries, reducing the environmental impact of downstream industry.

Nowadays, it is still difficult to imagine the global energetic matrix free of fossil transportation fuels, especially in developing economies, and raising the participation of renewable raw materials in crude oil refineries can be an attractive strategy. The Brazilian case reinforces that, even in nations with great demand for transportation fuels, biofuels can develop a fundamental role in the energetic matrix.

Although the recent forecasts indicate a falling demand for transportation fuel and a growing demand for petrochemicals, transportation fuel is still fundamental to sustaining the economic development of nations, especially in developing economies. This fact reinforces the necessity to reduce the carbon emissions even more in the crude oil processing chain, and biofuels can develop a fundamental role in the achievement of this goal in the downstream industry.

BIBLIOGRAPHY

1. The Catalyst Group. *Advances in Catalysis for Plastic Conversion to Hydrocarbons*, The Catalyst Group (TCGR), 2021.
2. Brazilian Petroleum Agency (ANP). *Brazilian Statistical Yearbook*, 2020.
3. Chang, R.J. Crude Oil to Chemicals—Industry Developments and Strategic Implications—Presented at Global Refining & Petrochemicals Congress (Houston, USA), 2018.
4. Charlesworth, R. Crude oil to Chemicals (COTC)—An Industry Game Changer? IHS Markit Company, Presented in 14th GPCA Forum, 2019.
5. Couch, K. The Refinery of the Future—A Flexible Approach to Petrochemicals Integration. Honeywell UOP Company, Presented in 12th Asian Downstream Summit, 2019.
6. Cui, K. *Why Crude to Chemicals is the Obvious Way Forward*, Wood Mackenzie, 2019.
7. Deloitte Company. *The Future of Petrochemicals: Growth Surrounded by Uncertainties*, 2019.
8. Energy Research Company (EPE). *Analysis of the Biofuels Conjuncture*, Technical Report, 2020.
9. Frecon, J., Le Bars, D., Rault, J. *Flexible Upgrading of Heavy Feedstocks*, PTQ Magazine, 2019.
10. Hilbert, T., Kalyanaraman, M., Novak, B., Gatt, J., Gooding, B., McCarthy, S. *Maximising Premium Distillate by Catalytic Dewaxing*, PTQ Magazine, 2011.

11. Lambert, N., Ogasawara, I., Abba, I., Redhwi, H., Santner, C. *HS-FCC for Propylene: Concept to Commercial Operation*, PTQ Magazine, 2014.
12. Maller, A., Gbordzoe, E. High Severity Fluidized Catalytic Cracking (HS-FCC™): From Concept to Commercialization, Technip Stone & Webster Technical Presentation to REFCOMM™, 2016.
13. Mukherjee, U., Gillis, D. *Advances in Residue Hydrocracking*, PTQ Magazine, 2018.
14. Muldoon, B.S. Profit Pivot Points in a Crude to Chemicals Integrated Complex—Presented at Ethylene Middle East Technology Conference, 2019.
15. Murphy, J.J., Payn, C.F. *Oil to Chemicals: New Approaches*, PTQ Magazine, 2019.
16. Refinery-Petrochemical Integration (Downstream SME Knowledge Share), Wood Mackenzie Presentation, 2019.
17. Reinventing the Refinery through the Energy Transition and Refining-Petrochemical Integration, IHS Markit, 2020.
18. Robinson, P.R., Hsu, C.S. *Handbook of Petroleum Technology*. 1st edition, Springer, 2017.
19. Sarin, A.K. Integrating Refinery with Petrochemicals: Advanced Technological Solutions for Synergy and Improved Profitability—Presented at Global Refining & Petrochemicals Congress (Mumbai, India), 2017.
20. Silva, M.W. *More Petrochemicals with Less Capital Spending*, PTQ Magazine, 2020.
21. Zhu, F., Hoehn, R., Thakkar, V., Yuh, E. *Hydroprocessing for Clean Energy—Design, Operation, and Optimization*. 1st edition, Wiley Press, 2017.
22. International Energy Agency (IEA). *Oil 2021: Analysis and Forecast to 2026*, 2021.
23. Silva, M.W., Clark, J. *Delayed Coking as a Sustainable Refinery Solution*, PTQ Magazine, 2021.

12 The Propylene Production Gap

The current scenario presents great challenges to the crude oil refining industry, such as the price volatility of raw materials, pressure from society to reduce environmental impacts, and increasingly lower refining margins. The newest threat to refiners is the reduction of the consumer market. In the last few years, it has become common news that countries intend to reduce or ban the production of vehicles powered by fossil fuels in the middle term, mainly in the European market. Despite the recent forecasts, the transportation fuel demand is still the main revenue driver for the downstream industry, as presented in Figure 12.1, based on data from Wood Mackenzie.

According to Figure 12.1, is expected a growing demand for petrochemicals, while transportation fuel tends to present falling consumption. Still, according to Wood Mackenzie data, presented in Figure 12.2, due to the higher added value, the most integrated refiners tend to achieve higher refining margins than the conventional refiners, which keep the operations focused on transportation fuels.

NCM = Net Cash Margins

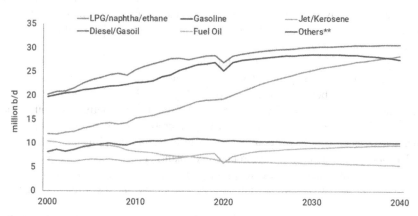

FIGURE 12.1 Global Oil Demand by Derivative

Source: Wood Mackenzie, 2020, with Permission

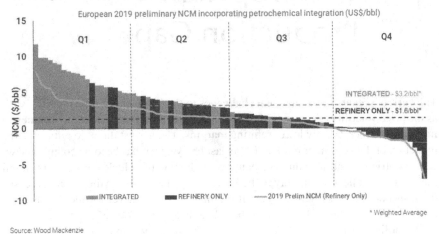

FIGURE 12.2 Refining Margins to Integrated and Non-integrated Refining Hardware

Source: Wood Mackenzie, 2020, with Permission

Despite the improvement in fuel efficiency, the growing market of electric vehicles decreased the participation of transportation fuel in the global crude oil demand. New technologies like additive manufacturing (3D printing) have the potential to produce a great impact on transportation demands, leading to even more impact on transportation fuel demand. Furthermore, the higher availability of lighter crude oil favors the oversupply of lighter derivatives that facilitate the production of petrochemicals against transportation fuels, as well as the higher added value of petrochemicals in comparison with fuels.

Facing these challenges, the search for alternatives that ensure the survival and sustainability of the refining industry became constant for refiners and technology developers. Due to their similarities, better integration between refining and petrochemical production processes appears as an attractive alternative. Although the advantages, it's important to consider that the integration between refining and petrochemical assets increases the complexity, requires capital spending, and affects the interdependency of refineries and petrochemical plants. These facts need to be deeply studied and analyzed case by case, and propylene maximization can be an attractive alternative to refiners to ensure participation in a profitable and high-demand market.

12.1 PROPYLENE: A FUNDAMENTAL PETROCHEMICAL INTERMEDIATE

Propylene is one of the most important petrochemical intermediates nowadays, being the second largest consumed in the world, behind ethylene. Propylene can be applied as an intermediate to the production of some fundamental products:

The Propylene Production Gap

- Acrylonitrile
- Propylene oxide
- Cumene
- Acrylic acid
- Polypropylene

Polypropylene is responsible for a major part of the propylene demand, followed by acrylonitrile and propylene oxide.

Propylene is normally produced in three commercial grades: Refinery grade has a purity varying from 50% to 70%. Refinery grade is applied in the production of cumene, for example. Chemical grade (92% to 96%) is applied to produce acrylonitrile, propylene oxide, and acrylic acid. Finally, polymer grade (up to 99,5%) is applied to produce polypropylene.

The main sources of propylene are the steam cracking process, FCC units in crude oil refineries, olefin metathesis, propane dehydrogenation, and methanol-to-olefin processes.

The steam cracking units are the main propylene supplier, followed by FCC units in crude oil refineries.

According to some recent forecasts, the petrochemical market tends to rise in the next years and, in the middle term, will be responsible for a major part of crude oil consumption more than transportation fuels. This fact has made refiners look for closer integration with petrochemical assets through the maximization of petrochemical intermediates in their refining hardware as a strategy to ensure better refining margins and higher value addition to crude oil.

According to 2019 data from Deloitte, a growth of 4,4% in ethylene demand and 4,1% in propylene demand is expected until 2022. Due to its higher added value and growing consumer market, the production of petrochemical intermediates has become the focus of many refiners and process technologies developers. This scenario is leading to a growing propylene production gap.

According to some forecasts, the demand for propylene will increase from 130 million metric tons in 2020 to around 190 million metric tons in 2030. Facing the increasingly light feed to refineries and steam cracking units, which tend to favor the ethylene production to the detriment of propylene, propylene demand tends to be supplied by on-purpose propylene production routes like propane dehydrogenation, methanol-to-olefin (MTO) processes, and olefin metathesis.

12.2 PROPYLENE PRODUCTION ROUTES

As quoted earlier, currently a major part of the propylene market is supplied by steam cracking units, but a great part of the global propylene demand is from the separation of LPG produced in FCC units. Figure 12.3 present a feedstocks and derivatives profile in a typical FCC unit.

Normally, the LPG produced in FCC units contains close to 30% of propylene, and the added value of the propylene is close to 2,5 times of the LPG. According to the local market, the installation of propylene separation units presents an attractive return on investment. Despite the advantage, a side effect of the propylene separation

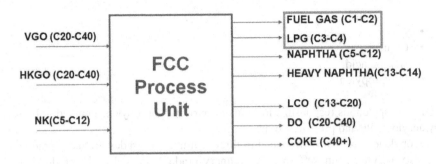

FIGURE 12.3 Production and Feedstocks Profile in a Typical FCC Unit

from LPG is that the fuel stays heavier, leading to specifications issues, mainly in colder regions. In these cases, alternatives are to segregate the butanes and send this stream to the gasoline pool, add propane to the LPG, or add LPG from natural gas. It's important to consider that some of these alternatives reduce the LPG offer, which can be a severe restriction according to the market demand.

A great challenge in the propylene production process is the propane and propylene separation step. The separation is generally hard by simple distillation because the relative volatility between propylene and propane is close to 1,1. This fact generally conducts to distillation columns with many equilibrium stages and high internal reflux flow rates.

There are two technologies normally employed in propylene-propane separation towers that are known as heat-pump and high-pressure configurations.

The high-pressure technology applies a traditional separation process that uses a condenser with cooling water to promote the condensation of top products. In this case, it's necessary to apply sufficient pressure to promote the condensation of products at the ambient temperature. Furthermore, the reboiler uses steam or another available hot source. The adoption of a high-pressure separation route requires a great availability of low-pressure steam in the refining hardware. In some cases, this can be a restrictive characteristic, and the heat pump configuration is more attractive, despite the higher capital requirements.

The separation process applying heat pump technology uses the heat supplied by the condensation of top products in the reboiler. In this case, the reboiler and the condenser are the same equipment. To compensate the non-idealities, it's necessary to install an auxiliary condenser with cooling water.

The application of heat pump technology allows a decrease in the operating pressure by close to 20 bar to 10 bar. This fact increases the relative volatility of propylene-propane, making the separation process easier and, consequently, reducing the number of equilibrium stages and internal reflux flow rate required for the separation.

Normally, when the separation process by distillation is hard (with relative volatilities lower than 1,5), the use of heat pump technology is more attractive.

Furthermore, some variables need to be considered during the choice of the best technology for the propylene separation process, like the availability of utilities, temperature gap in the column, and installation cost.

The Propylene Production Gap

Normally, propylene is produced in the refineries with specifications. The polymer grade is the most common and has higher added value with a purity of 99,5% (minimum). This grade is directed to the polypropylene market. The chemical grade where the purity varies between 90% and 95% is normally directed to other uses. A complete process flow diagram for a typical propylene separation unit applying heat pump configuration is presented in Figure 12.4.

The LPG from the FCC unit is pumped to a depropanizer column, where the light fraction (essentially a mixture of propane and propylene) is recovered at the top of the column and sent to a deethanizer column, while the bottom (butanes) is pumped to LPG or gasoline pool, according to the refining configuration. The top stream of the deethanizer column (lighter fraction) is sent back to the FCC, where it is incorporated into the refinery fuel gas pool or, in some cases, can be directed to petrochemical plants to recover the light olefins (mainly ethylene) present in the stream, while the bottom of the deethanizer column is pumped to the C_3 splitter column, where the separation of propane and propylene is carried out. The propane recovered in the bottom of the C_3 splitter is sent to the LPG pool, where the propylene is sent to the propylene storage park. The feed stream passes through a caustic wash treatment aiming to remove some contaminants that can lead to deleterious effects on petrochemical processes. An example is carbonyl sulfide (COS), which can be produced in the FCC (through the reaction between CO and S in the riser).

12.2.1 The Maximum Olefin Operation Mode

According to the market demand, FCC units can be optimized to produce the most demanded derivatives. Refiners facing gasoline surplus markets can operate the processing unit in maximum olefin operation mode to minimize the production of cracked naphtha.

In this operation mode, the FCC unit operates under high severity, translated to high operation temperature (TRX) and high catalyst/oil ratio. The catalyst formulation considers higher catalyst activity through the addition of ZSM-5 zeolite. There is the possibility of a reduction in the total processing capacity due to the limitations in blowers and cold area capacity.

An improvement is observed in the octane number of cracked naphtha despite a lower yield due to the higher aromatic concentration in the cracked naphtha. In some cases, the refiner can use the recycled cracked naphtha to improve the LPG yield even more.

In the maximum LPG operation mode, the main restrictions are the cold area processing capacity, metallurgic limits in the hot section of the unit, treating section processing capacity, and the top systems of the main fractionating column. In markets with falling demand for transportation fuels, this is the most common FCC operation mode.

By changing the reaction severity, it is possible to maximize the production of petrochemical intermediates, mainly propylene, in conventional FCC units.

The use of FCC catalyst additives, such as ZSM-5, can increase unit propylene production by up to 9%. Despite the higher operating costs, the higher revenues from the higher added value of derivatives should lead to a positive financial result for the

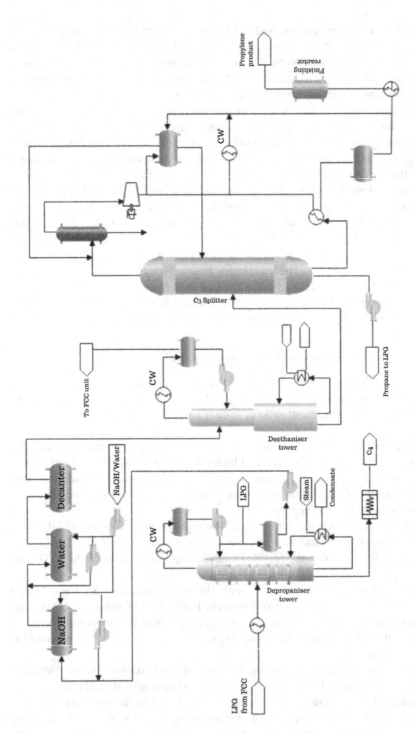

FIGURE 12.4 Typical Process Flow Diagram for an FCC Propylene Separation Unit Applying Heat Pump Configuration

The Propylene Production Gap

refiner, according to current market projections. A relatively common strategy also applied to improve the yield of LPG and propylene in FCC units is the recycling of cracked naphtha, leading to an over-cracking of the gasoline-range molecules.

Nowadays, the falling demand for transportation fuel has made the refiners optimize FCC units, aiming to maximize the propylene yield, following the trend of closer integration with the petrochemical sector. Among the alternatives to maximize the propylene yield in FCC units is the use of ZSM-5 as an additive to the FCC catalyst, as well as the adjustment of the process variables to most severe conditions, including higher temperatures and catalyst circulation rates. Another interesting alternative is to recycle the cracked naphtha to the processing unit aiming to improve the LPG and, consequently, the propylene yield.

The installation of propylene separation units can present a significant capital investment to refiners, but considering the last forecasts that reinforce the trend of reduction in the demand for transportation fuels, this investment can be a strategic decision for all players in the downstream industry in the middle term to ensure higher added value to the processed crude oil and market share.

12.2.2 The Petrochemical FCC Alternative

In markets with high demand for petrochemicals, the petrochemical FCC can be an attractive alternative to refiners aiming to ensure higher added value to bottom barrel streams. An example of FCC technology developed to maximize the production of petrochemical intermediates is the PetroFCC process, by UOP. This process combines a petrochemical FCC and separation processes optimized to produce raw materials for the petrochemical process plants. Other available technologies are the HS-FCC process, commercialized by Axens, and the Indmax process, licensed by Lummus.

For petrochemical FCC units, the reaction temperature reaches 600°C, and a higher catalyst circulation rate raises the gas production, which requires a scaling up of the gas separation section. The higher thermal demand makes the operation of the catalyst regenerator advantageous in total combustion mode, leading to the necessity of installing a catalyst cooler system.

In petrochemical FCC units, the combination of higher reaction temperature (TRX) and a cat/oil ratio five times higher when are compared to the conventional processing units and the petrochemical FCC can lead to a growth of the light olefin yield (ethylene + propylene + C_4s) from 14% to 40%. Figure 12.5 presents a schematic process for the Flexene technology developed by Axens with the purpose of maximizing propylene from FCC.

The installation of petrochemical catalytic cracking units requires a deep economic study considering the high capital investment and higher operational costs. However, some forecasts indicate growth of 4% per year in the market of petrochemical intermediates until 2025. This scenario can attract capital investment to raise the market share in the petrochemical sector, allowing the refiner a favorable competitive positioning through the maximization of petrochemical intermediates. Figure 12.6 presents a block diagram showing a case study demonstrating how the petrochemical

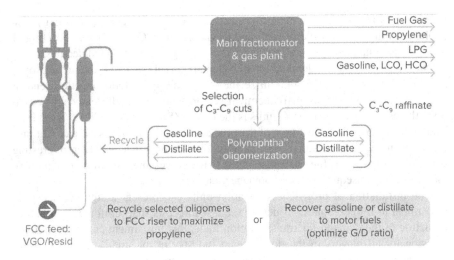

FIGURE 12.5 Schematic Process Flow for Flexene Technology by Axens (with Permission)

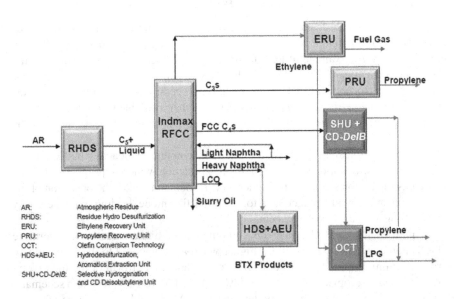

FIGURE 12.6 Olefin Maximization in the Refining Hardware with Indmax FCC Technology, by Lummus (with Permission)

FCC unit—in this case, the Indmax technology, by Lummus—can maximize the yield of petrochemicals in the refining hardware.

In refining hardware with conventional FCC units further than the higher temperature and catalyst circulation rates, it's possible to apply the addition of catalyst additives like the zeolitic material ZSM-5, which can raise the olefin yield close to 9% in some cases when compared with the original catalyst. This alternative raises operational costs. However, as aforementioned, it can be economically attractive, considering the petrochemical market forecasts.

Among other petrochemical FCC technologies, it's possible to quote the Maxofin process developed by KBR and the SCC technology, developed by Lummus.

12.2.3 Steam Cracking Units

The steam cracking process has a fundamental role in the petrochemical industry. Nowadays, most of the light olefins, light ethylenes, and propylenes are produced through the steam cracking route. Steam cracking consists of a thermal cracking process that can use gas or naphtha to produce olefins.

Naphtha to steam cracking is composed basically of straight-run naphtha from crude oil distillation units. Normally, to meet the requirements as petrochemical naphtha, the stream needs to present high paraffin content (higher than 66%).

Due to their relevance, great technology developers have dedicated their efforts to improving steam cracking technologies over the years, especially related to steam cracking furnaces. Companies like Stone & Webster, Lummus, KBR, Linde, and Technip develop technologies for the steam cracking process. One of the most known steam cracking technologies is the SRT process (Short Residence Time), developed by Lummus, which uses a reduced residence time to minimize the coking process and ensure a higher operational life cycle. Another commercial technology dedicated to optimizing the yield of ethylene is the Score technology, developed by KBR and ExxonMobil, which combines a selective steam cracking furnace with a high-performance olefin recovery section.

Cracking reactions occur in the furnace tubes. The main concern and limitation to the operating life cycle of steam cracking units is coke formation in the furnace tubes. The reactions are carried out under high temperatures, between 500°C to 700°C, according to the characteristics of the feed. For heavier feeds like gas oil, a lower temperature is applied to minimize coke formation. The combination of high temperatures and low residence time is the main characteristic of the steam cracking process. The focus of a naphtha steam cracking unit is normally producing ethylene, but the yield of propylene in a typical naphtha steam cracking unit can reach 15%.

12.2.4 Propane Dehydrogenation

Among the main technological routes dedicated to the production of these compounds, we can highlight the production of light olefins through the dehydrogenation route of light paraffin (C_2–C_5). According to the local market supplied by the refiner, the capital investment in processing units capable of producing light olefins through paraffin dehydrogenation can be an attractive strategy.

Light paraffin is normally commercialized as LPG or gasoline and presents reduced added value when compared with light olefins.

The dehydrogenation process involves the hydrogen removed from the paraffinic molecule and consequently hydrogen production, according to the reaction (12.1):

$$R_2CH\text{-}CHR_2 \leftrightarrow R_2C=CR_2 + H_2 \quad (1)$$

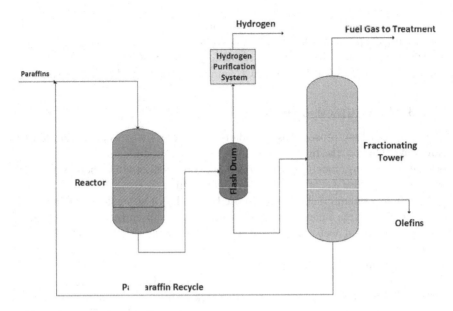

FIGURE 12.7 Process Flow Diagram for a Typical Light Paraffin Dehydrogenation Processing Unit

The dehydrogenation reactions have strongly endothermic characteristics, and the reaction conditions include high temperatures (close to 600°C) and mild operating pressures (close to 5 bar). The catalyst normally applied in the dehydrogenation reactions is based on platinum carried on alumina (other active metals can be applied).

Figure 12.7 shows a schematic process flow diagram for a typical dehydrogenation processing unit.

The main processes that can produce streams rich in light paraffin are physical separation processes such as LPG from atmospheric distillation and units dedicated to separating gases from crude oil.

The feed stream is mixed with the recycle stream before to entre to the reactor, the products are separated into fractionating columns, and the produced hydrogen is sent to purification units (normally PSA units) and posteriorly sent to consumers units as hydrotreating and hydrocracking, according to refining scheme adopted by the refiner. Light compounds are directed to the refinery or petrochemical complex fuel gas pool after adequate treatment, while the olefinic stream is directed to the petrochemical intermediate consumer market.

During the dehydrogenation process, there is a strong tendency for coke deposition on the catalyst surface, and the regeneration of the catalytic bed is periodically carried out through controlled combustion of the produced coke. Some process arrangements present two reactors in parallel aim to optimize the processing unit's operational availability. In these cases, while a reactor is in production, the other is in the regeneration step.

The Propylene Production Gap

Due to the growing market and high added value of light olefins, great technology developers have dedicated their efforts to developing paraffin dehydrogenation technologies. UOP developed and commercializes the Oleflex, which is capable of producing olefins from paraffin dehydrogenation with a continuous catalyst regeneration process. Despite the higher initial investment, this technology can minimize the unavailability period to regenerate the catalyst.

Another available technology is the Catofin process, licensed by Lummus. As aforementioned, in this case, two reactors are used in parallel, as presented in Figure 12.8.

Among the dehydrogenation technologies available, we can note the STAR process, commercialized by ThyssenKrupp-Uhde, and the FBD process, by SnamProgetti.

Due to their chemical characteristics, olefinic compounds can be employed in the production of a large number of interesting products such as polymers (polyethylene and polypropylene), propylene oxide, and oxygenated compound production intermediates (MTBE, ETBE, etc.).

As a process of high energy consumption, there is a great variety of research in the sense of developing more active and selective catalysts that reduce the need for energetic contribution to the dehydrogenation process. One of the main variations of the dehydrogenation process is the process called oxidative dehydrogenation, which occurs according to Reaction 12.2.

$$R_2CH\text{-}CHR_2 + O_2 \leftrightarrow R_2C\text{=}CR_2 + H_2O \tag{12.2}$$

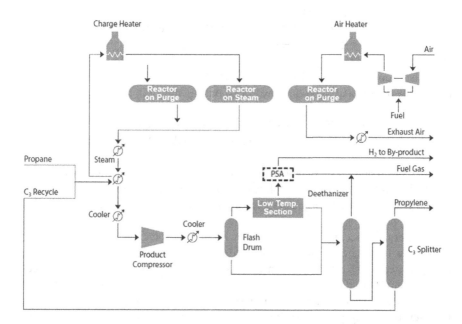

FIGURE 12.8 Simplified Process Scheme to Catofin Dehydrogenation Technology, by Lummus (with Permission)

This reaction is strongly exothermic, and this is the main advantage in relation to the traditional dehydrogenation process due to the high risk of paraffin combustion against the dehydrogenation reaction.

12.2.5 Olefin Metathesis

The olefin metathesis process involves the combination of ethylene and butene to produce propylene, as presented in Reaction 12.3.

$$H_2C=CH_2 + H_3C\text{-}HC=CH\text{-}CH_3 \rightarrow 2\ H_2C=HC\text{-}CH_3 \qquad (12.3)$$

The main technology licensors for olefin metathesis processes are Lummus and IFP (Institut Français du Pétrole). Figure 12.9 presents a basic process flow arrangement for the OCT technology, developed by Lummus.

The economic viability of olefin metathesis units relies on the price gap between propylene and ethylene, as well as the ethane availability in the market.

12.2.6 Methanol-to-Olefin Technologies (MTO)

Another alternative route to producing liquid hydrocarbons from syngas is the non-catalytic conversion of natural gas to methanol, followed by the polymerization to produce alkenes. Methanol is produced from natural gas according to the following chemical reactions:

$$CH_4 + H_2O = CO + 3H_2 \text{ (Steam Reforming)}$$
$$CO + H_2O = CO_2 + H_2 \text{ (Shift reaction)}$$
$$2H_2 + CO = CH_3OH \text{ (Methanol Synthesis)}$$

In the sequence, the methanol is dehydrated to produce dimethyl ether, which is posteriorly dehydrated to produce hydrocarbons, as shown in this sequence:

$$2\ CH_3OH = CH_3OCH_3 + H_2O \text{ (Methanol Dehydration)}$$
$$CH_3OCH_3 = C_2H_4 + H_2O \text{ (Dimethyl Ether Dehydration)}$$

The conversion of methanol to olefins or hydrocarbons is called methanol-to-olefin (MTO) or methanol-to-gasoline (MTG) technologies. Figure 12.10 presents a typical unit dedicated to producing methanol from natural gas through the two-step reforming process.

An alternative technology developed by Haldor Topsoe to produce methanol from natural gas is the altothermal reforming process, called ATR, which offers improvements related to the reforming furnace. A significant advantage of the ATR process is the lower required ratio of steam/carbon in the reforming step, which offers a great scale economy when compared with traditional production processes (one-step and two-step reforming processes). Figure 12.11 presents a basic process flow for the ATR process, developed by Haldor Topsoe.

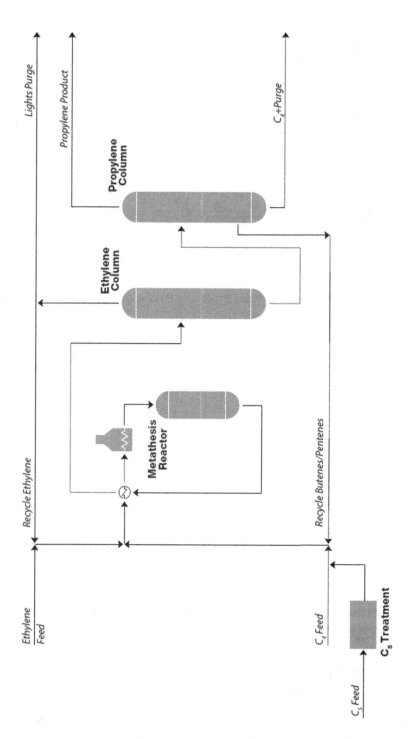

FIGURE 12.9 Process Flow Diagram for OCT Olefin Metathesis Technology, by Lummus (with Permission)

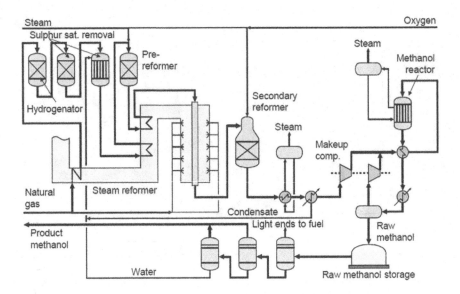

FIGURE 12.10 Haldor Topsoe Methanol Production Process from Natural Gas through Two-Reforming Process (with Permission)

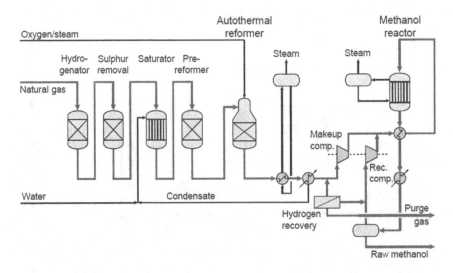

FIGURE 12.11 Altothermal (ATR) Process for Methanol Production from Natural Gas, by Haldor Topsoe (with Permission)

The most known processes dedicated to converting methanol into hydrocarbons are the MTG process, developed by ExxonMobil, and the MTO-Hydro process, developed by UOP.

MTO technologies present some advantages in relation to Fischer-Tropsch processes, as they show higher selectivity in hydrocarbon production. Furthermore, the obtained products require lower additional processing steps to achieve commercial specifications. Another important point is that the installation cost is normally lower than MTO process plants when compared with Fischer-Tropsch units, as these units are economically viable only on a large scale. Regarding the olefin production, the maximization of these derivatives can be especially attractive in the current scenario where there is a trend of reduction in transportation fuel demand, followed by the growing market of petrochemicals, creating the necessity for closer integration between refining and petrochemical assets to maximize the added value, share risks and costs, and ensure market share in a highly competitive scenario of the downstream sector. Other great technology developers for the methanol production process are Johnson Matthey, Linde, Chiyoda, Jacobs, and Lurgi.

The propylene maximization in the refining hardware can offer attractive opportunities to refiners, especially those inserted in markets with saturated demand for transportation fuel, like gasoline.

Nowadays, it is still difficult to imagine the global energetic matrix free of fossil transportation fuels, especially in developing economies. Despite this fact, recent forecasts and growing demand for petrochemicals, as well as the pressure to minimize the environmental impact produced by fossil fuels, create a positive scenario and act as the main driving force to the closer integration between refining and petrochemical assets. In extreme scenarios, zero-fuel refineries tend to grow in the middle term, especially in developed economies.

The synergy between refining and petrochemical processes raises the availability of raw material to petrochemical plants and makes the supply of energy to these processes more reliable while ensuring better refining margin to refiners due to the high added value of petrochemical intermediates when compared with transportation fuels. The development of crude-to-chemicals technologies reinforces the necessity of closer integration of refining and petrochemical assets by the brownfield refineries aiming to face the new market that tends to be focused on petrochemicals instead of transportation fuels. It's important to note the competitive advantage of the refiners from the Middle East that have easy access to light crude oil, which can be easily applied in crude-to-chemicals refineries.

As presented previously, closer integration between refining and petrochemical assets demands high capital spending. Despite this fact, the installation of refining units capable of adding value to the naphtha can be a significant competitive advantage among the refiners, especially those players inserted in the market with a gasoline surplus. The advantage of the naphtha to chemicals routes can be exemplified through the growing "propylene gap" present in the text, where the refiners are capable of maximizing the propylene yield both for on-purpose or traditional production (FCC of steam cracking) and can enjoy a significant competitive advantage in the market.

Although petrochemical integration has benefits, it's fundamental to keep in mind the necessity to reach a circular economy in the downstream industry. To achieve this goal, the chemical recycling of plastics is essential. As presented previously, there are promising technologies that can ensure the closing of the sustainability cycle of the petrochemical industry.

BIBLIOGRAPHY

1. Gary, J.H., Handwerk, G.E. *Petroleum Refining: Technology and Economics*. 4th edition, Marcel Dekker, 2001.
2. Marsh, M., Wery, J. *On-Purpose Propylene Production*, PTQ Magazine, 2019.
3. Mukherjee, M., Vadhri, V., Revellon, L. *Step-Out Propane Dehydrogenation Technology for the 21st Century*, The Catalyst Review, 2021.
4. Myers, R.A. *Handbook of Petroleum Refining Processes*. 3rd edition, McGraw-Hill, 2004.
5. Peiretti, A. *Haldor Topsoe—Catalyzing Your Business*, Technical Presentation of Haldor Topsoe Company, 2013.
6. Wood Mackenzie Company. *Refinery-Chemicals Integration: How to Benchmark Success*, Wood Mackenzie Presentation, 2020.
7. Wood Mackenzie Company. *Refinery-Petrochemical Integration (Downstream SME Knowledge Share)*, Wood Mackenzie Presentation, 2019.
8. Sawyer, G. *Basics of the Chemical Industry- Propylene and Its Products*, AIChE, 2014.

13 Gas-to-Liquid Processing Routes

Despite the current scenario where is observed a surplus of crude oil followed by low prices, the downstream industry constantly lives with uncertainties related to the guarantee of access and supply of crude oil in the quantity and quality required. Problems related to geopolitics and even the exhaustion of existent recoverable reserves are driving forces for the development of alternative technologies for crude oil. This topic is especially attractive to countries that present a lack of a significant amount of crude oil reserves and great dependence on crude oil derivatives, such as Japan and China.

In this case, the look for alternatives to crude oil aiming to sustain the energy demand necessary to sustain economic development become a strategic issue, as mentioned earlier, mainly considering the volatility of crude oil prices and the geopolitical scenario. Despite the higher cost in comparison with the traditional crude oil refining, the necessity of the production of high-quality fuels can make the nonconventional alternatives attractive. An example was the coal liquids hydrogenation during the Second World War to produce liquid fuels on a large scale due to the scarcity of crude oil in some countries. Crude oil production players with great reserves of natural gas can find gas-to-liquid technologies an attractive way to ensure higher added value to their natural resources.

13.1 GAS-TO-LIQUID TECHNOLOGIES

One of the most promising and well-developed technologies currently is the conversion of syngas ($CO + H_2$) into longer-chain hydrocarbons, such as gasoline and other liquid fuel products, known as gas-to-liquid (GTL) technologies. Liquid hydrocarbon production can be carried out by direct syngas conversion, in Fischer-Tropsch synthesis reactions, or through methanol production as an intermediate product (methanol-to-olefin technologies).

Fischer-Tropsch is a chemical process through is possible the liquid hydrocarbon production according to the following chemical reactions:

$$\text{Paraffin Production: } n\,CO + (2n+1)H_2 = C_nH_{2n+2} + nH_2O$$
$$\text{Olefin Production: } n\,CO + 2nH_2 = C_nH_{2n} + nH_2O$$

These reactions are strongly exothermic, and the CO/H_2 ratio in the syngas is a key parameter in defining the hydrocarbon chain extension that will be produced.

The reactions occur normally under temperatures that vary from 200°C to 350°C and operating pressures in the range of 15–30 bar. The catalyst commonly applied to these reactions is based on cobalt or iron as active metals deposited upon alumina as a carrier.

Figure 13.1 presents a block diagram for a typical process plant dedicated to producing liquid hydrocarbons from Fischer-Tropsch synthesis.

The process in Figure 13.1 is based on the syngas gas generation from steam reforming of natural gas. This is the most common route. However, there are process variations applying syngas production through coal, biomass, or petroleum coke gasification route.

The process starts with syngas generation and, as aforementioned, the produced hydrocarbon chain extension is controlled in the Fischer-Tropsch synthesis step through the CO/H_2 ratio in the syngas fed to the Fischer-Tropsch reactors (beyond temperature and reaction pressure), then the produced hydrocarbons are separated and sent to refining steps such as isomerization, hydrotreating, hydrocracking, and catalytic reforming, according to the application of the produced derivative (gasoline, diesel, lubricant, etc.).

Some side reactions can occur during the hydrocarbon production process, leading to coke deposition on the catalyst, causing its deactivation according to the following chemical reactions:

$$2CO = C + CO_2 \text{ (Boudouard Reaction)}$$
$$CO + H_2 = C + H_2O \text{ (CO Reduction)}$$

The type of reactor used in the Fischer-Tropsch synthesis has a strong influence on the yield and quality of the obtained products. The campaign time of the processing units also depends on the type of reactor. Fixed-bed reactors are widely employed

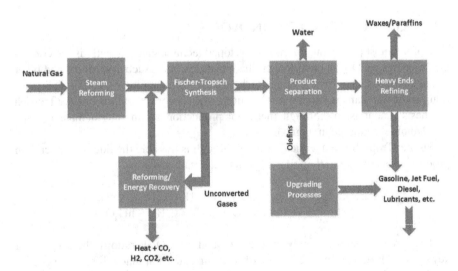

FIGURE 13.1 Block Diagram of a Typical Fischer-Tropsch GTL Process Plant

Gas-to-Liquid Processing Routes

in the Fischer-Tropsch synthesis. However, there is a reduced campaign time due to the low resistance to catalyst deactivation phenomenon. Modern processing units use fluidized bed or slurry phase reactors that present a higher resistance to coke deposition on the catalyst and better heat distribution, leading to higher campaign periods.

Most recently, a reduction trend in demand for transportation fuel has been observed, and some refiners are looking to change their production focus from transportation fuel to petrochemicals. Gas-to-liquid technology can be applied in synergy with conventional refining processes to improve the yield of petrochemicals in the refining hardware through the production of high-quality naphtha that can be applied to FCC or steam cracking units to produce light olefins, ensuring higher added value to the processed crudes and gas as well as participation in a growing market.

Another attractive alternative and synergy opportunity for refiners is the production of ammonia, which is the base of any fertilizer. Despite the flat demand over the last few years, a growing market is expected in the next years due to the increasing demand for food at a global level. According to recent forecasts, presented in Figure 13.2, growing demand for methanol is also expected in the next years. This intermediate can be used to produce high-demand products, like formaldehyde, which is applied to produce plastics and coatings, allowing great added value to the crude oil and natural producing chain.

It's interesting to note that methanol keeps its price despite the economic crisis due to the COVID-19 pandemic. This fact reinforces that chemical production has an edge as refiners and petrochemical producers face volatility in the downstream sector.

13.1.1 AVAILABLE TECHNOLOGIES

Several geopolitical crises in history motivated the development of new technologies and the improvement of the original Fischer-Tropsch process. There is a wide array of process technologies developed aiming at liquid hydrocarbon production from syngas. Among the principal available technologies we can note are the Synthol and SPD process, developed by Sasol; the Gasel process, by Axens; the E-Gas technology, by ConocoPhillips; the SCGP, by Shell; and TIGAS, developed by Haldor Topsoe; among others.

Figure 13.3 presents a basic process flow diagram for the Gasel technology, developed by Axens, which applies a slurry phase reactor.

As quoted earlier, one of the main advantages of GTL technologies is the possibility of using several raw materials to produce syngas, which ensures great flexibility. In regions with large coal availability, the gasification technologies have strategic character given the great restriction of this fuel use in the energetic matrix due to the high environmental impact. In these cases, the coal conversion into syngas and posteriorly in liquid hydrocarbons is very economically attractive. Another alternative is to apply renewable raw material (biomass) to produce syngas.

On the other hand, in regions with great availability and easy access to large natural gas reserves, syngas production through natural gas reforming steam still is

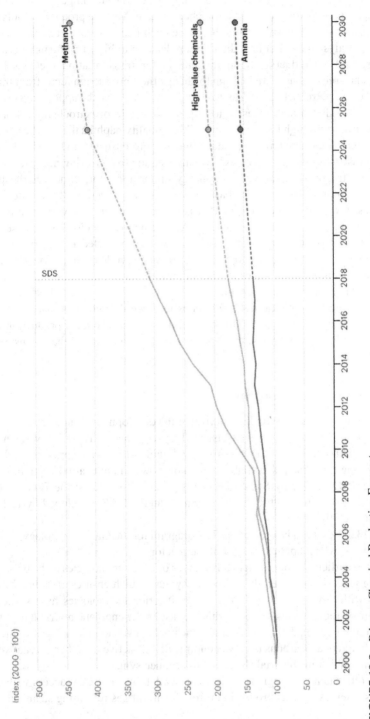

FIGURE 13.2 Primary Chemical Production Forecast
Source: IEA (2020)

Gas-to-Liquid Processing Routes

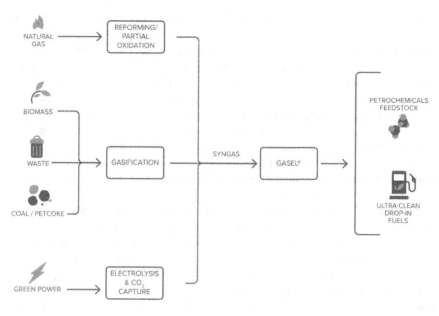

FIGURE 13.3 Process Flow Diagram for Gasel Technology, by Axens (with Permission)

shown as the most economical route to produce this raw material on an industrial scale.

An alternative route to producing liquid hydrocarbons from syngas is the non-catalytic conversion of natural gas to methanol, followed by polymerization to produce alkenes. Methanol is produced from natural gas according to the following chemical reactions:

$$CH_4 + H_2O = CO + 3H_2 \text{ (Steam Reforming)}$$
$$CO + H_2O = CO_2 + H_2 \text{ (Shift reaction)}$$
$$2H_2 + CO = CH_3OH \text{ (Methanol Synthesis)}$$

In the sequence, methanol is dehydrated to produce dimethyl ether, which is posteriorly dehydrated to produce hydrocarbons, as shown in the sequence:

$$2 CH_3OH = CH_3OCH_3 + H_2O \text{ (Methanol Dehydration)}$$
$$CH_3OCH_3 = C_2H_4 + H_2O \text{ (Dimethyl Ether Dehydration)}$$

The methanol conversion to olefins into hydrocarbons is called methanol-to-olefin (MTO) or methanol-to-gasoline (MTG) technologies. The most known processes dedicated to converting methanol in hydrocarbons are the processes MTG developed by ExxonMobil Company and the MTO-Hydro process, developed by UOP.

MTO technologies present some advantages in relation to Fischer-Tropsch processes, as they show higher selectivity in hydrocarbon production. Furthermore, the obtained products require lower additional processing steps to achieve commercial specifications. Another important point is that the installation cost is normally lower than MTO process plants when compared with Fischer-Tropsch units, as these units are economically viable only on a large scale. Regarding the olefin production, the maximization of these derivatives can be especially attractive in the current scenario where there is a trend of reduction in transportation fuel demand, followed by the growing market of petrochemicals, creating the necessity for closer integration between refining and petrochemical assets to maximize the added value, share risks and costs, and ensure market share in a highly competitive scenario of the downstream sector.

The streams produced by GTL technologies show reduced contaminant content (sulfur, nitrogen, etc.) and high quality, which makes these products attractive from the point of view of environmental footprint and profitability. These characteristics lead to lower operating costs when compared with the conventional crude oil refining route as a lower hydroprocessing capacity is needed and less severe processes can be a significant competitive differential. In scenarios with crude oil shortage or overpricing, these technologies can be competitive and strategic in some nations.

Despite these advantages, the streams produced by the Fischer-Tropsch process tend to present a linear chain (essentially normal paraffin) that leads to poor cold flow properties. In market consumers with cold weather, the use of hydroprocessing units with dewaxing beds can be necessary to meet the quality regulations and ensure adequate performance of the final derivatives. Aiming to minimize this issue, some researchers are studying an upgrade for the traditional Fischer-Tropsch process using a new catalyst. The catalyst Pt/H-ZSM-22 zeolite appears like the most promising, aiming to isomerize selectively the n-paraffin produced by the Fischer-Tropsch process, improving the quality and the added value of the produced streams. It is important to consider the high cost of the platinum catalyst, a necessary and adequate economic evaluation to support the decision to change the catalyst.

13.2 AMMONIA PRODUCTION PROCESS: AN OVERVIEW

The ammonia production process is one of the most important advances of science in the last century as it allows human development through the minimization of concerns related to the capacity to produce foods on a large scale.

The ammonia production process can be chemically described through the following chemical reaction:

$$N_2 + 3\,H_2 \leftrightarrow 2\,NH_3 \tag{13.1}$$

Reaction 13.1 is slightly exothermic and normally is applied a catalyst based on iron (Fe), which is normally filled in the reactor in oxide form (Fe_2O_3) and activated to metallic form in a reaction with hydrogen.

Over time, some additives were developed to improve the activity and stability of the catalyst. Some examples are TiO_2, SiO_2, V_2O_5, and CaO. Nowadays, commercial catalysts based on iron are considered very stable, reaching an operation life cycle of close to ten years.

Gas-to-Liquid Processing Routes

The ammonia production process is favored by higher pressures, the commercial processes apply operating pressures varying from 140 to 300 kgf/cm². Due to the exothermic characteristic of Reaction 13.1, the process is favored by lower temperatures, but to meet the catalyst limitations and kinetic restrictions, relatively high temperatures are applied (higher than 400°C).

The reaction presents low conversion per pass in the reactor, and this fact tends to accumulate N_2 and H_2 in the reaction system disfavoring the chemical equilibrium. To solve this, ammonia is recovered from the reactor effluent through refrigeration, and the N_2 and H_2 are recycled in the reactor. The presence of NH_3 in the recycler also disfavors the chemical equilibrium, and it's necessary to control the NH_3 quantity in the recycle stream, normally below that 5%.

Another common issue in ammonia plants is the inert accumulation in the reaction system (methane and argon). This is controlled through a constant purge of the recycled gas after the refrigeration system.

The main application of ammonia nowadays is to produce nitrogen-based fertilizers like urea.

13.2.1 Ammonia Production Technologies: Some Commercial Processes

Due to their relevance to society and high added value, the ammonia production process has been a target of a lot of research through the years, and some important technology developers and academies have dedicated their efforts to improving the traditional Haber-Bosch process.

Companies like BASF, KBR, Thyssen-Krupp, and Haldor Topsoe (among others) are well known as specialists in the development of ammonia production processes at a commercial level.

The main improvement looked at by the technology developers is related to more efficient ammonia recovery systems and contaminant removal processes like CO_2 removal through amine treating, and higher conversion per pass in the reactor, always looking to reduce the energy demand in the process. Figure 13.4 presents an ammonia production plant designed by Haldor Topsoe, where a side-fired reformer is applied to achieve high efficiency in the reforming section.

As aforementioned, the demand for ammonia tends to grow in the next years, and the capital investment in ammonia production plants can ensure higher revenues and competitiveness for refiners, especially for those inserted in markets with a high offer of natural gas. This can be especially attractive to refiners with low-complexity refining hardware, which can present a high difficult to reach closer integration with petrochemical assets, following recent trends, due to the high capital spending required.

Nowadays, one of the main concerns of the players in the downstream industry is related to more sustainable routes to produce hydrogen than the traditional methane steam reforming, which presents a high generation of carbon dioxide (CO_2). Ammonia has been considered an attractive alternative as a hydrogen carrier as it is relatively easily liquified under moderate pressures (1 mpa at 25°C). These characteristics can lead the users to oversupply one of the main issues of hydrogen that is the low volumetric energy density. The ammonia can be stored in common pressure vessels

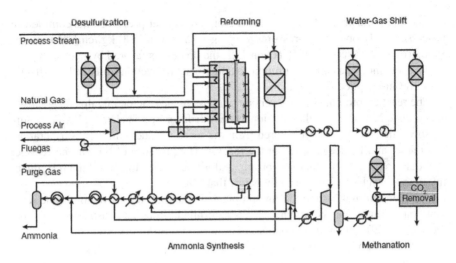

FIGURE 13.4 Ammonia Production Plant Designed by Haldor Topsoe (with Permission)

and have about 17,65% in mass of hydrogen and can be relatively easily converted to hydrogen through ammonia decomposition reaction over an adequate catalyst. According to the literature, the ammonia decomposition reaction can achieve an adiabatic efficiency of 85%. Despite the limitations like the tradeoff between the cost operations of higher temperatures and costs to remove unconverted ammonia, these routes can be applied to transport hydrogen produced by clean routes over long distances, especially considering the tradeoff, which the refiners are involved nowadays where there are increasing necessity for hydrogen and growing pressure to reduce the emissions of greenhouse gases, putting pressure over the traditional hydrogen production route through natural gas steam reforming.

Another characteristic that can make the GTL technologies even more attractive in the next years is the increasing restriction on CO_2 emissions from fossil fuel combustion; among them is natural gas. The possibility of fixing carbon through a GTL process can represent an efficient and profitable alternative.

The environmental footprint is one of the great concerns of society related to the crude oil production chain, and the use of GTL technologies can minimize the greenhouse gases, allowing a more sustainable and profitable growth in society.

As shown earlier, GTL technologies can represent an attractive alternative to some nations aiming to ensure a source of high-quality liquid hydrocarbons capable of sustaining economic development in a scenario of lack of crude oil resources and supply crisis and allow a higher value addition to their natural resources.

The high demand for chemicals and petrochemicals, accompanied by the reduction in transportation fuel demand at a global level, is leading refiners to look for creative alternatives to ensure relevance and market share in growing markets to sustain their economic results.

Ammonia production can be an attractive way to achieve higher revenues and participation in a growing market while diversifying the refining business, according to the recent trends of the downstream sector, especially as a renewable hydrogen carrier.

BIBLIOGRAPHY

1. Gary, J.H., Handwerk, G.E. *Petroleum Refining: Technology and Economics*. 4th edition, Marcel Dekker, 2001.
2. IEA (International Energy Agency). *Primary Chemical Production in the Sustainable Development Scenario, 2000–2030*, 2020.
3. Pattabathula, V., Richardson, J. *Introduction to Ammonia Production*, American Institute of Chemical Engineers (AIChE), 2016.
4. Robinson, P.R., Hsu, C.S. *Handbook of Petroleum Technology*. 1st edition, Springer, 2017.
5. S&P Global Platts. *Petrochemical Trends H1 2021- Demand Recovery Possible Despite Ongoing Uncertainty*, 2021.
6. Silva, M.W. *An Alternative to Crude Oil*, Hydrocarbon Engineering Magazine, 2020.

14 Business Strategy Models Applied to the Downstream Industry

In 1979, Michel Porter wrote the revolutionary article in *Harvard Business Review*, "How Competitive Forces Shapes Strategy," where he introduced the concept of the five competitive forces. Through an analysis of these forces in a determined business, the players can analyze their competitive positioning while it is possible to define some strategies to achieve better competitive positioning.

According to Michel Porter's article, there are five competitive forces that define the competitive positioning of a player in a determined market:

- Supplier power: What is the bargaining power of the supplier in relation to the consumers?
- Customer power: How flexible is the customer and what are the customer's alternatives in relation to your services and products?
- Substitute products and services: Are there substitute products or services capable of easily substituting the products/services currently offered?
- Threat of new entrants: How difficult is it for a new entrant to join the market?
- Rivalry between the existing players: How aggressively are the players competing in the market?

Figure 14.1 presents the relation of the five competitive forces for a determined market.

According to the positioning of a player in relation to each one of the competitive forces, the players can define their strategies to improve the competitive positioning reinforcing the point considered the weakness.

The current scenario presents great challenges to the crude oil refining industry, such as the price volatility of raw materials, pressure from society to reduce environmental impact, and increasingly lower refining margins. The drastic reduction of sulfur content in the final product leads refiners to look for alternatives to reduce the sulfur content in the intermediate streams. In this business environment, it's possible to imagine how Porter's competitive forces affect the downstream industry.

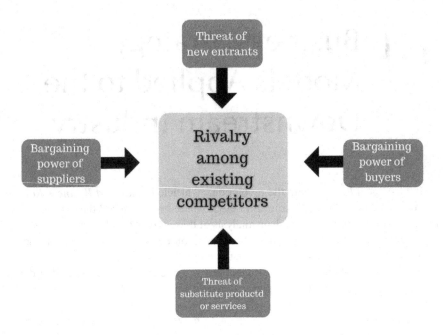

FIGURE 14.1 The Five Competitive Forces
Source: Modified from Porter, M., 2008

14.1 PORTER'S COMPETITIVE FORCES IN THE DOWNSTREAM INDUSTRY

Considering what is shown in Figure 14.1, it's possible to analyze the five competitive forces listed by Michael Porter in the downstream industry.

- Bargaining power of suppliers: The main supplier of the downstream industry is the crude oil producers. Normally, the refiners have low bargaining power because the crude oil price is defined by several factors, but refiners relying on flexible refining hardware can face advantages as they are capable of processing heavier and discounted crudes that present lower costs. In other words, adequate bottom barrel conversion capacity can offer a significant competitive advantage to the refiners. Over the years, some companies have developed integrated operations to minimize the exposition of the variation of crude oil prices. Regarding the other suppliers, normally, the refiners are considered great customers, and these suppliers tend to present low bargaining power. In normal conditions, they do not represent a great threat. In this case, the most integrated players can get a competitive advantage. Operational efficiency is another fundamental characteristic. Refiners capable of reducing the operating costs can acquire more resilience faced with

the variations in crude oil prices. The operating cost reduction is especially related to the energy efficiency of the refining hardware, as more than 60% of the operating costs are related to energy consumption.
- Bargaining power of buyers: The customers have low bargaining power in the downstream industry as it is still difficult to find energy sources in quantity and quality capable of substituting crude oil derivatives. Of course, in markets with a high quantity of players, competitiveness can offer alternatives to the customers, but it's difficult to achieve a great gap in prices in a commodity market. Despite this, public opinion over the downstream industry is increasingly important and has the potential to change the energy market. An example is the growing trend of energy transition efforts demanded by society, requiring a transition to low-carbon energy sources.
- Threat of new entrants: Due to the high capital requirements, it's hard to face the new entrant threat in the downstream industry, but this threat can always be considered mainly due to government interventions and the attractiveness of the local markets.
- Rivalry among existing competitors: This is a great concern in the downstream industry. The great number of players and the standardization of the products create great pressure on the refining margins. To overcome this, refiners have looked to improve their operational efficiency, but it's normally quickly followed by the other players, reducing the profitability in the market.
- Threat of substitute products and services: Nowadays, this is a great threat to the players of the downstream industry. The reduction of the consumer market in the last few years has become common. Many countries intend to reduce or ban the production of vehicles powered by fossil fuels in the middle term, mainly in the European market. Despite the recent forecasts, the transportation fuel demand is still the main revenue driver for the downstream industry, as presented in Figure 14.2, based on data from Wood Mackenzie.

According to Figure 14.2, the transportation fuel demand represents close to five times the demand by petrochemicals and a focus on transportation fuel of the current refining hardware, considering the data from 2019. Despite these data, a trend of stabilization in transportation fuel demand is observed close to 2030, followed by a growing market of petrochemicals. Still, according to Wood Mackenzie's data, presented in Figure 14.3, a relevant growth is expected in the petrochemical participation in the global oil demand.

The improvement in fuel efficiency and the growing market of electric vehicles tend to decrease the participation of transportation fuel in the global crude oil demand. Figure 14.4 present the growth of electric vehicles in the last few years in the global market.

Other than the electrification of the automobiles, new technologies like additive manufacturing (3D printing) have the potential to produce a great impact on transportation demands, leading to even more impact on transportation fuel demand. The

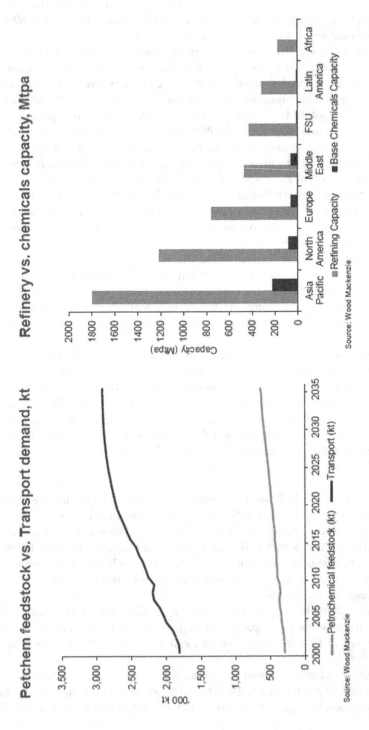

FIGURE 14.2 Relation to Petrochemical Feedstock/Transportation Fuel Feedstock and Installed Capacity

Source: Wood Mackenzie (2019)

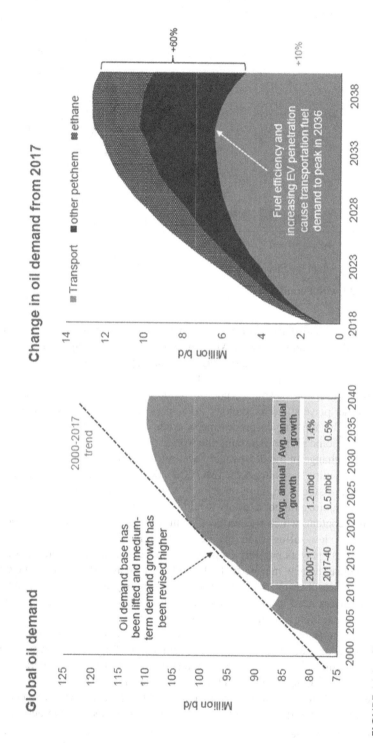

FIGURE 14.3 Change in the Profile of Global Crude Oil Demand

Source: Wood Mackenzie (2019)

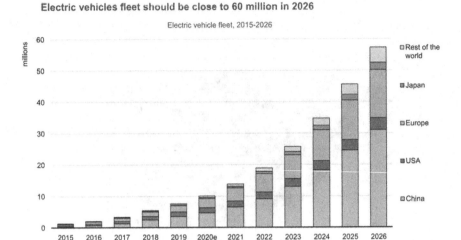

FIGURE 14.4 Growth of the Electric Vehicles Fleet over the Years

Source: IEA, 2020, with Permission

growing trend of vehicle-sharing services, like Uber, has great potential to destroy demand in the downstream industry. Another threat is the growing participation of renewable raw materials in crude oil refineries, in response to society's requirement for energy transition efforts. In the last months, some important players have announced the conversion of some crude oil refineries into renewable processing plants, while other players and technology developers announced the production of diesel and jet fuel by applying co-processing of crude oil and renewable raw materials like HVGO in some refineries around the world.

Another great change in the downstream sector that reinforces the necessity of a high conversion refining hardware is the IMO 2020. Restrictive regulations like IMO 2020 raised the pressure even more on refiners with low bottom barrel conversion capacity as they require a higher capacity to add value to residual streams, especially related to sulfur content, which was reduced from 3,5% (in mass) to 0,5%. Refiners with easy access to low-sulfur crude oils present a relative competitive advantage in this scenario. These players can rely on relatively low-cost residue upgrading technologies to produce the new marine fuel oil (bunker), such as carbon rejection technologies (solvent deasphalting, delayed coking, etc.), but they are the minority in the market. Most players need to look for sources of low-sulfur crudes, which present higher costs, putting their refining margins under pressure, or look for deep bottom barrel conversion technologies to ensure more value addition to processed crude oil and avoid losing competitiveness in the downstream market. For these refiners, deepest residue upgrading like hydrocracking technologies can offer great operational flexibility, despite the high capital spending. In this scenario, with the necessity for higher value addition to the bottom barrel stream and the growing market of petrochemicals, refiners with adequate bottom barrel conversion capacity can achieve a great competitive advantage in the downstream industry.

Business Strategy Models Applied to the Downstream Industry 235

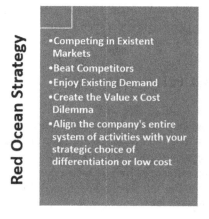
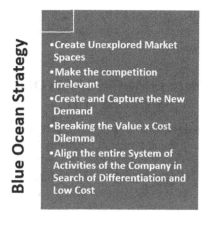

FIGURE 14.5 Differences between Blue and Red Ocean Strategies
Source: Modified from Kim & Mauborgne, 2004

Based on the earlier description, it's possible to apply the article published by W. Chan Kim and Renée Mauborge, called "Blue Ocean Strategy," in *Harvard Business Review*, to classify the competitive markets in the downstream industry. In this article, the authors define the conventional market as a red ocean where the players tend to compete in the existing market, focusing on defeating competitors through the exploration of existing demand, leading to low differentiation and low profitability. The blue ocean is characterized by looking for a space in non-explored (or few explored markets), creating and developing new demands, and reaching differentiation. This model can be applied (with some specificities once is a commodity market) to the downstream industry, considering the traditional transportation fuel refineries and the petrochemical sector.

Due to its characteristics, the transportation fuel market can be imagined as the red ocean, where the margins tend to be low and under high competition between the players with low differentiation capacity. On the other side, the petrochemical sector can be faced like the blue ocean where few players are able to meet the market in competitive conditions, higher refining margins, and significant differentiation in relation to refiners dedicated to transportation fuel market. Figure 14.5 presents the basic concept of the blue ocean strategy in comparison with the traditional red ocean, where the players fight for market share with low margins.

As presented previously, the market forecasts indicate that the refiners able to maximize petrochemicals against transportation fuel can achieve highlighted economic performance in the short term. In this sense, crude-to-chemicals technologies can offer an even more competitive advantage to the refiners with a capacity for capital investment.

It can be difficult to some people to understand the term "differentiation" in the downstream industry once this is a market that deals with commodities, but the differentiation here is related to the capacity to reach more added value to the processed crude oil, and as presented earlier, nowadays this is translated in the capacity to maximize the petrochemical yield, creating differentiation between integrated and non-integrated players.

14.2 CHANGING THE FOCUS: MORE PETROCHEMICALS AND LESS FUEL

In this business environment, it's possible to adapt the Ansoff matrix to consider the contraction profile of the transportation fuel market to analyze the available alternatives to the downstream players. The Ansoff matrix is presented in Figure 14.6.

In Figure 14.6, the current position of downstream players is focused on transportation fuel demand that presents a contraction profile as aforementioned. In this scenario, there are three alternatives to the players:

1. Look for new clients: This alternative seems attractive at first look, but the stricter regulations and trend of reduction in the consumption create great pressure on the consumption of fossil fuels. The major consumers of transportation fuel are still the in developing economies, like Brazil, Mexico, and India, but the most efficient engines and substitute technologies, like hybrid and electric vehicles, tend to reduce the market growth even in these countries.
2. Offer a new value addition: Face the reduction in transportation fuels, an attractive strategy to the downstream sector is to offer a new proposed value to the market through higher value addition to the processed crude oil, as well as needed materials to society with a lower environmental footprint than fossil fuels. Petrochemical intermediates have higher added value to

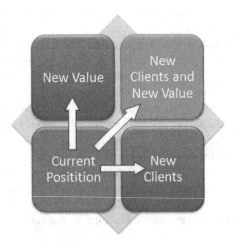

FIGURE 14.6 Adapted Ansoff Matrix to Contraction Market

Source: Based on Rogers, 2016

refiners and growing demand, as aforementioned. The substitution of steel is some engineering materials is an interesting way to secure the market for petrochemicals in the short term. In this sense, refiners can change the production focus from transportation fuel to petrochemicals, especially in markets like Asia and Europe, where the fall in transportation fuel demand is most significant. Beyond petrochemicals, the capacity to add value to bottom barrels streams appears like a competitive advantage.

3. New clients and new value addition: Strategically, this alternative seems the right way to follow, mainly for refiners with the most complex refining hardware. Through the promotion of closer integration with the petrochemical sector, refiners not only offer a higher proposed value to the clients and society but can reach a new range of customers capable of ensuring higher added value to the processed crude oil and lower operational costs through available synergies between refining and petrochemical assets.

14.2.1 Petrochemical and Refining Integration as a Differentiation Strategy

The focus of the closer integration between refining and petrochemical industries is to promote and seize the synergies and existing opportunities between both downstream sectors to generate value for the whole crude oil production chain. Table 14.1 presents the main characteristics of the refining and petrochemical industry and the synergies' potential.

As aforementioned, the petrochemical industry has been growing at considerably higher rates when compared with the transportation fuel market in the last few years. Additionally, the industry represents a noble destiny and is less environmentally aggressive than crude oil derivatives. The technological bases of the refining and petrochemical industries are similar, which leads to possibilities of synergies capable of reducing operational costs and add value to derivatives produced in the refineries.

TABLE 14.1
Refining and Petrochemical Industry Characteristics

Refining Industry	Petrochemical Industry
Large feedstock flexibility	Raw material from naphtha/NGL
High capacities	Higher operation margins
Self-sufficient in power/steam	High electricity consumption
High hydrogen consumption	High availability of hydrogen
Streams with low added value (unsaturated gases and C_2)	Streams with low added value (heavy aromatics, pyrolysis gasoline, C_4)
Strict regulations (e.g., benzene in gasoline)	Strict specifications (hard separation processes)
Transportation fuel demand declining at a global level	High-demand products

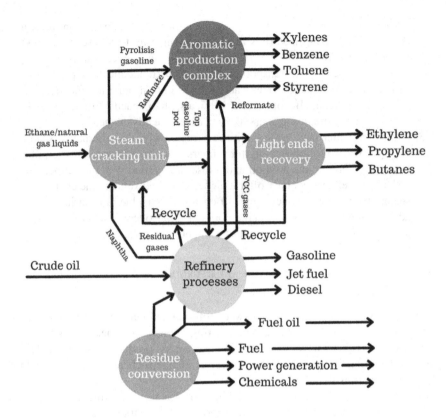

FIGURE 14.7 Synergies between Refining and Petrochemical Processes

Figure 14.7 presents a block diagram that shows some integration possibilities between refining processes and the petrochemical industry.

Process streams considered with low added value to refiners like fuel gas (C_2) are attractive raw materials to the petrochemical industry, as well as streams considered residual to petrochemical industries (butanes, pyrolysis gasoline, and heavy aromatics) can be applied to refiners to produce high-quality transportation fuels, this can help the refining industry meet the environmental and quality regulations to derivatives.

The integration potential and the synergy among the processes rely on the refining scheme adopted by the refinery and the consumer market. Processing units such as FCC and catalytic reforming can be optimized to produce petrochemical intermediates to the detriment of streams that will be incorporated into the fuel pool. In the case of FCC, the installation of units dedicated to producing petrochemical intermediates, called petrochemical FCC, to reducing to the minimum the generation of streams to produce transportation fuels. However, the capital investment is high once the severity of the process requires the use of material with the noblest metallurgical characteristics.

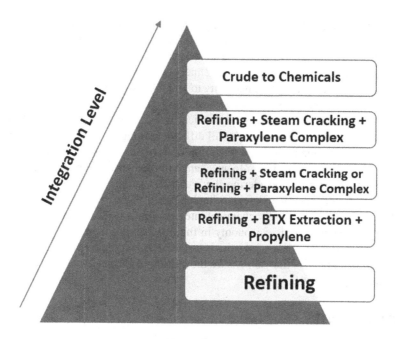

FIGURE 14.8 Petrochemical Integration Levels
Source: Modified from IHS Markit, 2018

IHS Markit proposed a classification of the petrochemical integration grades, as presented in Figure 14.8.

According to the classification proposed, the crude-to-chemicals refineries are considered the maximum level of petrochemical integration, where the processed crude oil is totally converted into petrochemical intermediates, like ethylene, propylene, and BTX.

Nowadays, it is still difficult to imagine the global energetic matrix free of fossil transportation fuels, especially in developing economies. Despite this fact, recent forecasts and growing demand for petrochemicals, as well as the pressure to minimize the environmental impact produced by fossil fuels, create a positive scenario and act as the main driving force to the closer integration between refining and petrochemical assets. In extreme scenarios, zero-fuel refineries tend to grow in the middle term, especially in developed economies.

The synergy between refining and petrochemical processes raises the availability of raw material to petrochemical plants and makes the supply of energy to these processes more reliable while ensuring better refining margin to refiners due to the high added value of petrochemical intermediates when compared with transportation fuels. The development of crude-to-chemicals technologies reinforces the necessity of closer integration of refining and petrochemical assets by the brownfield refineries aiming to face the new market that tends to be focused on petrochemicals instead of transportation fuels. It's important to note the competitive advantage of the refiners from the Middle East that have easy access to light crude oil, which can be easily applied in crude-to-chemicals refineries. As presented previously, crude-to-chemicals refineries are based on deep conversion processes that require high capital spending. This fact can put the refiners under pressure with restricted access to capital, again reinforcing the necessity to look for close integration with the petrochemical sector to achieve competitiveness.

On the extreme side of the petrochemical integration trend, there are zero-fuel refineries. As quoted previously, it's still difficult to imagine the downstream market without transportation fuels, but it seems a serious trend, and the players of the downstream sector need to consider a focus change in their strategic plans, like opportunity and threat, mainly considering the pressure on transportation fuel due to the decarbonization necessity and new technologies.

Despite the benefits of petrochemical integration, it's fundamental to consider the necessity to reach a circular economy in the downstream industry. To achieve this goal, the chemical recycling of plastics is essential. As presented previously, there are promising technologies that can ensure the closing of the sustainability cycle of the petrochemical industry.

BIBLIOGRAPHY

1. The Catalyst Group. *Advances in Catalysis for Plastic Conversion to Hydrocarbons*, The Catalyst Group (TCGR), 2021.
2. Chang, R.J. Crude Oil to Chemicals—Industry Developments and Strategic Implications—Presented at Global Refining & Petrochemicals Congress (Houston, USA), 2018.
3. Cui, K. *Why Crude to Chemicals Is the Obvious Way Forward*, Wood Mackenzie, 2019.
4. Frecon, J., Le Bars, D., Rault, J. *Flexible Upgrading of Heavy Feedstocks*, PTQ Magazine, 2019.
5. Gary, J.H., Handwerk, G.E. *Petroleum Refining: Technology and Economics*. 4th edition, Marcel Dekker, 2001.
6. Gupta, K., Aggarwal, I., Ethakota, M. *SMR for Fuel Cell Grade Hydrogen*, PTQ Magazine, 2020.
7. Kim, W.C., Mauborge, R. *Blue Ocean Strategy*, Harvard Business Review, 2004.
8. Mukherjee, U., Gillis, D. *Advances in Residue Hydrocracking*, PTQ Magazine, 2018.
9. Porter, M.E. *The Five Competitive Forces that Shape Strategy*, Harvard Business Review, 2008.
10. Wood Mackenzie Company. *Refinery-Petrochemical Integration (Downstream SME Knowledge Share)*, Wood Mackenzie Presentation, 2019.
11. Rogers, D.L. *The Digital Transformation Playbook: Rethink your Business for the Digital Age*. 1st edition, Columbia University Press, 2016.
12. Sarin, A.K. Integrating Refinery with Petrochemicals: Advanced Technological Solutions for Synergy and Improved Profitability—Presented at Global Refining & Petrochemicals Congress (Mumbai, India), 2017.
13. Silva, M.W. *More Petrochemicals with Less Capital Spending*, PTQ Magazine, 2020.
14. Vu, T., Ritchie, J. *Naphtha Complex Optimization for Petrochemical Production*, UOP Company, 2019.
15. International Energy Agency (IEA). *Oil 2021: Analysis and Forecast to 2026*, 2021.

15 Corrosion Management in Refining Assets

The highly competitive environment of the refining industry requires high availability and reliability of the refining hardware in the sense of maximizing the operational life cycle of the processing units and avoiding unnecessary shutdowns and production losses. One of the great threats to the availability and integrity of equipment in the refining industry is the corrosion phenomena, which can lead to a reduction in the operation life cycle of process equipment and, in extreme cases, serious accidents.

Among the corrosion mechanisms found the crude oil refining industry is naphthenic corrosion. This corrosion mechanism occurs in hot sections of the processing units like crude oil distillation and delayed coking units. Naphthenic corrosion leads to quickly material loss, representing a dangerous threat to the integrity of these processing units.

15.1 NAPHTHENIC CORROSION: GENERAL OVERVIEW

The naphthenic corrosion occurs in processing units that process bottom barrel streams that tend to present a high concentration of naphthenic acids and operate under high temperatures. Due to these characteristics, the naphthenic corrosion phenomenon is observed in crude oil distillation units and residue upgrading units like delayed coking.

The characteristics of the processed crude oil slate are a determinant factor in naphthenic corrosion. A very relevant characteristic of oils for refining hardware is naphthenic acidity. Naphthenic acidity is determined based on the amount of KOH required to neutralize 1 gram of crude oil. Normally, a mixture of crude oils is sought in the refinery load so that it does not exceed 0,5 mg KOH/g. Above this reference, the bottom sections of the distillation units can undergo a severe corrosive process, leading to shorter periods of the operational campaign and higher operating costs in addition to problems associated with integrity and safety. Naphthenic acidity is directly linked to the concentration of oxygenated compounds in crude oil that tend to be concentrated in the heavier fractions, giving instability and odor to the intermediate currents.

The sulfur content in crude oil is another key factor in the naphthenic corrosion phenomena. In crude oil with a sulfur content higher than 2%, a protective layer of iron sulfide (FeS) is formed on the metal surface that is insoluble, preventing the attack by naphthenic acids, as presented in Figure 15.1. In this sense, refiners processing very low-sulfur crude oil with high total acid number (TAN) can face severe issues with naphthenic corrosion in their refining hardware.

DOI: 10.1201/9781003291824-15

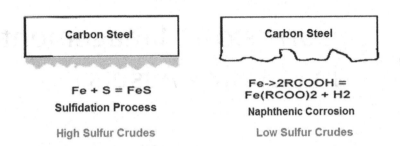

FIGURE 15.1 Naphthenic Corrosion Process

Equation 15.1 presents the chemical representation of naphthenic corrosion.

$$Fe + 2RCOOH \rightarrow Fe(RCOO)_2 + H_2 \qquad (15.1)$$

According to the literature, above 400°C, the iron naphthenate (corrosion product) is soluble in hydrocarbons and is attacked by H_2S, leading to the regeneration of naphthenic acid, as represented in Equation 15.2.

$$Fe(RCOO)_2 + H_2S \rightarrow FeS + 2RCOOH \qquad (15.2)$$

In crude oil distillation units, the bottom sections of the atmospheric column and the vacuum distillation column are the most common regions where naphthenic corrosion is observed, while in delayed coking units, the phenomenon is observed in the bottom section of the main fractionator column.

The carbon steel, series 300 and 400 stainless steel, and nickel alloys tend to undergo naphthenic corrosion.

Among the actions to control the naphthenic corrosion in the refining hardware is the blending of crude oils to keep the TAN and sulfur content inside the adequate limits. Other alternatives are the injection of neutralizers or corrosion inhibitors in the processing streams and the selection of materials with higher resistance to naphthenic acid attack.

Refineries processing crudes with high acidity and low sulfur normally use chromium and molybdenum alloys in the bottom sections of crude oil distillation columns and transfer pipes, as well as in residue upgrading units. In extreme conditions, it's possible to use stainless steel 317 L, which presents a high resistance to naphthenic corrosion. This decision needs to consider the higher capital investment due to the high cost of this material.

The availability of the refining hardware is a key parameter to ensure the economic sustainability of the refiners, especially those inserted in highly competitive markets.

As described previously, naphthenic corrosion can compromise the reliability and availability of the key units to the refining hardware. For this reason, naphthenic corrosion issues may need to consider crude oil selection to minimize integrity risks and shorter operational campaigns. In this sense, adequate monitoring and control of the corrosion process in these units are fundamental to ensuring the competitiveness of the players in the downstream sector.

15.2 CORROSION IN SOUR WATER STRIPPING UNITS

To keep the environmental under control impacts in the crude oil refining sector, some processing technologies were developed over the years and installed in the refining hardware. Nowadays, it's impossible to think in the downstream sector without the environmental processes units due to the current environmental requirements, and the performance and reliability of these units are fundamental to the refiners' strategy to achieve the most profitable and cleaner operations. Considering the current regulations and the necessity to reduce the environmental impact of the crude oil refining industry, ensuring the reliability and availability of environmental processing units is fundamental to the players of the downstream industry, and the corrosion processes are a great threat to this objective. Due to its characteristics and process conditions, the corrosion phenomenon in sour water stripping units is a special concern in the crude oil refining industry.

The corrosion process in sour water stripping units normally occurs through hydrogen attack and salt deposition (ammonium bisulfide and ammonium chloride). The main regions under corrosion attack are the heating system of bottom columns and top sections, as presented in Figure 15.2.

The feed systems of the sour water stripping units tend to experience corrosion by hydrogen attack. The hydrogen attack mechanism involves the contamination of steel by hydrogen (H_2), leading to the risk of failure, especially in periods of instability, such as stopping and starting the processing unit. Reaction 15.3 presents the atomic hydrogen formation in regions with the presence of H_2S.

$$Fe + H_2S \rightarrow FeS + 2H \tag{15.3}$$

The main failure mechanisms provoked by hydrogen attack are the cracking induced by hydrogen, carbon steel embrittlement due to molecular hydrogen formation, and stress corrosion induced by hydrogen attack.

The main corrosion process observed in the sour water stripping units is related to ammonium salt deposition (bisulfide and chloride). The corrosion process resulting from the salt deposition is associated with the contaminants present in the charge currents of the sour water units. The processing of currents containing sulfur and nitrogen leads to the formation of H_2S and NH_3 in the unit's outlet currents as recycled gases. Such gases can be combined to produce ammonium bisulfide (NH_4HS), according to Reaction 15.4.

$$NH_{3(g)} + H_2S_{(g)} \leftrightarrow NH_4HS_{(s)} \tag{15.4}$$

The concentration of salts formed depends on the content of contaminants in the unit's load, so sour water from units dedicated to processing bottom barrel streams tends to show more severe processes of corrosion by this mechanism.

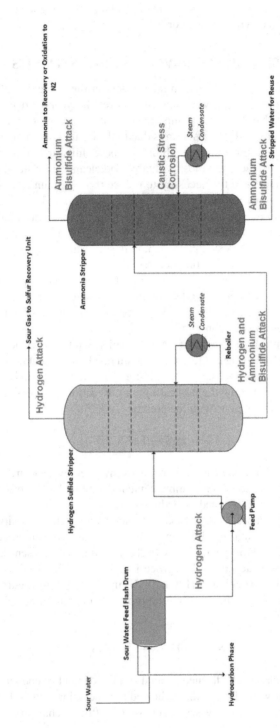

FIGURE 15.2 Predominant Corrosion Mechanisms in Sour Water Stripping Units

The ammonium chloride formation process is similar, as presented in Reaction 15.5.

$$NH_{3(g)} + HCl_{(g)} \leftrightarrow NH_4Cl_{(s)} \qquad (15.5)$$

The salt deposition leads basically to two corrosion processes in the units. In regions with low flow speed, corrosion under salt deposits is observed, while in regions with high flow speed, the most significant corrosion process is the corrosion-erosion mechanism. Due to these characteristics, the salt concentration tends to occur at the top of stripping towers, as presented in Figure 15.2.

NH_4HS tends to accumulate and provoke corrosion after the top condensers, while the ammonium chloride tends to undergo deposition at the top of the columns, as presented in Figure 15.2.

Among the actions to keep under control the corrosion process associated with ammonium salt deposition, the design of the processing units needs to consider the flow speed in the systems. A recommended range is 3,5–7 m/s. The injection of water wash is another key parameter to ensure salt removal and corrosion prevention. The water needs to be injected upstream of top condensers aiming to reduce the salt concentration.

To avoid salt deposition in sour water stripping units, especially NH_4HS, the NH_4HS precipitation curve must be considered.

In the great part of the sour water stripping units, the deposition temperature of NH_4HS is between 45 and 80°C, and the operating temperature of the stripping sections is controlled above these values to avoid NH_4HS deposition. The corrosion under deposits tends to occur with salt concentration higher than 2% in mass and with low flow speed (<3,5 m/s), while the corrosion-erosion process tends to occur with high flow speed (>7 m/s).

In the heating section of NH_3 stripping tower, it's common to observe the caustic stress corrosion process. This phenomenon tends to occur in welded connections that do not undergo thermal treatment for stress relief. The main susceptible materials are carbon steel and stainless steel (300 series). Some refiners tend to inject a caustic solution to achieve adequate pH for NH_3 stripping (close to 10). This practice needs to consider the risks of failures related to caustic stress corrosion, and the design of the processing units needs to consider the thermal treatment for the stress relief of the welded connections at the bottom section of NH_3 stripping columns.

Despite the efforts over the years, the crude oil refining industry still presents a great environmental impact, and crude oil derivatives are fundamental to sustaining economic development. In this sense, the availability of the environmental units is fundamental to allowing sustainable operation of the refining hardware from economic and environmental points of view. The unavailability of environmental processing units, like sour water stripping, can lead to the reduction in the processing capacity of the refining hardware aiming to keep under control the atmospheric emissions and leading to great economic losses and the risk of a shortage of crude oil derivatives in the market in extreme cases, and the corrosion processes are among the main threats to the reliability and availability of the sour water stripping units. It's always important to consider the relevance of the environmental units to the refining hardware, and the optimization and maintenance priorities of these units need to be put at the same level as processing units.

15.3 CORROSION PROCESS IN AMINE TREATING UNITS

The corrosion process in sour water stripping units normally occurs through degraded amine attack in the circuit of lean amine due to contaminant concentration (H_2S and CO_2) in the bottom section of the absorber column and hydrogen attack due to the presence of H_2S and cyanide (and thermally stable salts) in the top section of regenerator column. The main regions under corrosion attack are the heating system of bottom columns and top sections, as presented in Figure 15.3.

The hot section of amine treating units tends to undergo corrosion attacks due to degraded amine. During the treatment, thermal and chemical degradation of amines is common, producing compounds that reduce the absorption capacity of the amine solution and can lead to foam formation that raises the corrosion rates. The amine degradation process also raises the particulate concentration, which can improve the probability of the corrosion-erosion process to occur.

The main amine degradation process is due to the reaction with CO_2, as presented in Equation 15.6.

$$CO_2 + 2R_2NH \leftrightarrow 2R_2NH_2^+ + R_2NCO_2^- \qquad (15.6)$$

The amine degradation process is strongly dependent on the CO_2 partial pressure.

In refining hardware with FCC units, a common corrosion issue in amine treating units is related to the chemical degradation of amines through the reaction with formic acid, as presented in Equation 15.7.

$$HCOOH + (HOC_2H_4)_2NH \rightarrow (HOC_2H_4)_2NCOH + H_2O \qquad (15.7)$$

The formic acid formation in FCC units is common due to the reaction of CO and hydroxyl ions (OH-), as well as HCN hydrolysis.

Among the actions to minimize the corrosion process due to degraded amine, the most important is to keep an adequate amine makeup rate to avoid the concentration of degraded amine in the circuit. The amine filtration system is another key factor in ensuring the reliability of the unit as it is capable of removing degradation products and keeping under control the solid concentration in the lean amine.

The operating variables are another key factor. In DEA treating units, the temperature in the bottom section of the regeneration column needs to be kept below 130°C to minimize the amine degradation effect, and the steam to the bottom reboiler should present a maximum temperature of 140°C. The ratio of sour gases and free amine (pickup ratio) needs to be controlled; according to the literature, it should be between 0,25 and 0,5 in MEA treating units and between 0,5 and 0,8 in DEA treating units.

Corrosion Management in Refining Assets

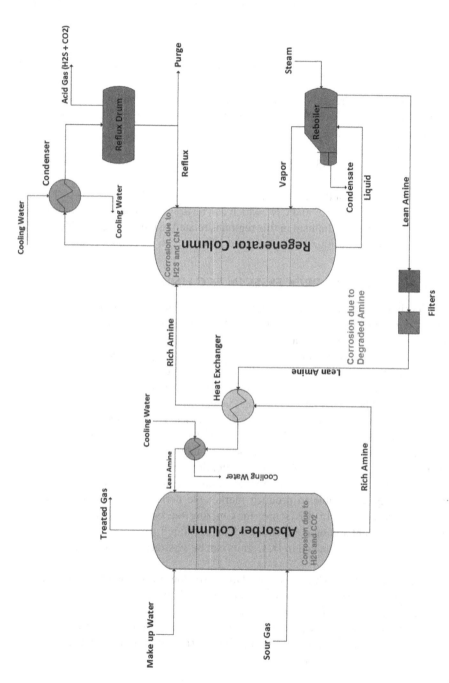

FIGURE 15.3 Predominant Corrosion Mechanisms in Amine Treating Units

The main corrosion process observed in the bottom section of the absorber column is related to the H$_2$S and CO$_2$ attack under low pH values. The main chemical reactions are presented here:

$$Fe + 2CO_2 + 2H_2O \leftrightarrow Fe_2+ + 2HCO_3- + 2H \quad (15.8)$$
$$Fe + H_2S \rightarrow FeS + 2H \quad (15.9)$$

Despite the occurrence of Equation 4, in an amine environment, the most common corrosion process related to H$_2$S in solution is the NH$_4$HS formation, as presented in Equation 15.10.

$$HS- + Fe + H_2O \leftrightarrow FeS + H_2 + OH- \quad (15.10)$$

In the top section of the regeneration tower, it's common to observe corrosion processes associated with H$_2$S and CN-, which can be dragged from FCC units. Another common corrosion mechanism in this region is related to the stress corrosion process due to amine concentration and hydrogen attack.

15.4 CORROSION PROCESSES IN FCC UNITS

Considering the current scenario and the forecasts, it is expected that FCC units greatly contribute to the economic sustainability of the downstream industry. In this sense, ensuring the higher availability of these units is fundamental to allow refiners to reach competitive operations, especially considering the operation under most severe conditions aiming to improve the yield of petrochemicals.

One of the great threats to the availability and integrity of equipment in the refining industry is the corrosion phenomena that can lead to a reduction in the operation life cycle of process equipment and, in extreme cases, grave accidents.

Some processing units, like FCC and hydroprocessing units, can undergo severe corrosion processes due to certain process conditions. The corrosion process in hydroprocessing units has been special attention in the last few years due to the strong dependence of the downstream sector on these technologies, considering that it's practically impossible to produce marketable derivatives without at least one hydroprocessing step. In this sense, maximizing the operation life cycle of these units can represent a great competitive advantage to refiners.

The corrosion phenomena in FCC units can be separated into two study areas:

1. Corrosion process in hot areas: This corrosion process is observed in the reactor section and can be carried out through erosion, oxidation, graphitization, refractory degradation, stress corrosion, and so on.
2. Corrosion process in cold areas: Phenomena observed in cold areas of FCC units, characterized by hydrogen attack, corrosion by ammonium salts, and corrosion-erosion process.

In this short technical note, we will present an overview of the corrosion process in cold areas of FCC units.

Corrosion Management in Refining Assets

The main corrosion process observed in the separation area of FCC units is related to ammonium salt deposition (bisulfide and chloride). The corrosion process resulting from the salt deposition is associated with the contaminants present in the charge currents of FCC units. The processing of currents containing sulfur and nitrogen leads to the formation of H_2S and NH_3 in the unit's outlet currents as recycled gases. Such gases can be combined to produce NH_4HS, according to Reaction 15.11.

$$NH_{3(g)} + H_2S_{(g)} \leftrightarrow NH_4HS_{(s)} \qquad (15.11)$$

The concentration of salts formed depends on the content of contaminants in the unit's load, so units dedicated to processing bottom barrel streams tend to show more severe processes of corrosion by this mechanism.

The ammonium chloride formation process is similar, as presented in Reaction 15.12.

$$NH_{3(g)} + HCl_{(g)} \leftrightarrow NH_4Cl_{(s)} \qquad (15.12)$$

The salt deposition leads basically to two corrosion processes in the units. In regions with low flow speed, corrosion under salt deposits is observed, while in regions with high flow speed, the most significant corrosion process is the corrosion-erosion mechanism. Due to these characteristics, the salt concentration tends to occur at the top of the main fractionators.

The ammonium bisulfide tends to accumulate and provoke corrosion after the top condensers, while the ammonium chloride tends to deposit at the top of the column.

Among the actions to keep under control the corrosion process associated with ammonium salt deposition, the design of the processing units needs to consider the flow speed in the systems. A recommended range is 3,5 to 7 m/s. The injection of water wash is another key parameter to ensure salt removal and corrosion prevention. Water needs to be injected upstream of top condensers.

To monitor the efficiency of water wash systems, refiners can measure the salt concentration in the sour water in the top drum and the pH in the sour water.

Regarding the damages produced by hydrogen, the phenomenon is related to the capacity of hydrogen to spread in the alloy structure. The attack by hydrogen can occur through atomic hydrogen dissolved in the metal, precipitation of molecular hydrogen, and hydrogen reaction with specific phases of the metal structure. The failures normally observed by hydrogen attack are cracking and/or blistering.

A key control parameter is the presence of free water and H_2S concentration. Normally, the concentration of H_2S higher than 50 ppm can favor the hydrogen attack, as well as the combination of low pH (lower than 4) with the presence of H_2S in the sour water. Cyanide (CN-) concentration is also a key parameter to hydrogen attack. Cyanide tends to attack the protection film, creating favorable conditions for hydrogen attack.

To avoid the corrosion processes related to hydrogen attack, the refiners can apply noble materials. In this case, it's necessary to carry out a balance between the capital investment and the operating costs to keep the reliable operation with conventional materials. Another alternative is to apply corrosion inhibitors in the key points of the processing units. In this case, the disadvantage is increasing operating costs.

The most applied strategy among the refiners is the control of the operating variables aiming to keep under control the corrosivity, allowing the use of carbon steel. Among the actions that can be carried out, we can note the following:

- H_2S removal, upstream of the compression section
- Minimizing water concentration in the system
- Controlling the pH between 7,5 and 8,2
- Converting cyanide to neutral forms, such as thiocyanate

Due to its simplicity, the most common action is continuous washing of the compression system. The quantity of water applied relies on the characteristics of the feed (contaminant content) and the capacity of top drums to remove sour water.

In any case, the corrosion monitoring needs to be a synergic work between the optimization and inspection teams, aiming to ensure reliable operations.

In summary, the basic recommendations to keep under control the corrosion phenomenon in FCC units are as follows:

1. Adequate design of the separation drums to allow the capacity to remove sour water without drag
2. Ensuring efficient water wash systems capable of ensuring low salt concentrations and pH in the range of 7,5–8,2
3. Keeping the cyanide concentration under control
4. Carrying out detailed inspections in maintenance shutdowns

As aforementioned, FCC units have a key role in the refining hardware, and the adequate availability is a great competitive differential in the downstream industry, especially considering the higher demand for petrochemicals in the near future, according to the recent forecasts.

15.4.1 The Petrochemical FCC: Raising Competitive Advantage x Corrosion Attention

Considering the current scenario of the downstream industry and the last forecasts, it's been an observed trend that there is a reduction in transportation fuel demand, followed by a growing market of petrochemicals, leading the refiners to optimize their FCC units to maximum LPG yield to improve the capacity to produce light olefins and promote closer integration with petrochemical assets. A major part of the catalytic cracking units is optimized to maximize transportation fuels, especially gasoline. However, faced with the current scenario, some units have been optimized to maximize the production of light olefins (ethylene, propylene, and butenes). As aforementioned, units focused on this goal have these operational conditions severely changed, raising the cracking rate.

The reaction temperature reaches 600°C, and a higher catalyst circulation rate raises the gas production, which requires a scaling up of the gas separation section.

The higher thermal demand makes it advantageous to operate the catalyst regenerator in total combustion mode, leading to the necessity of installing a catalyst cooler system.

The installation of a catalyst cooler system raises the processing unit profitability through the total conversion enhancement and selectivity to noble products, such as propylene and naphtha, against gases and coke production. The catalyst cooler is necessary when the unit is designed to operate under total combustion mode due to the higher heat release rate, as presented here:

$$C + \tfrac{1}{2} O_2 \rightarrow CO \text{ (Partial Combustion) } \Delta H = -27 \text{ kcal/mol}$$
$$C + O_2 \rightarrow CO_2 \text{ (Total Combustion) } \Delta H = -94 \text{ kcal/mol}$$

In this case, the temperature of the regeneration vessel can reach values close to 760°C, leading to higher risks of catalyst damage, which is minimized by installing a catalyst installation. Furthermore, the higher temperature in the regenerator requires materials with the noblest metallurgy. This significantly raises the installation costs of these units, which can be prohibitive to some refiners with restricted capital access. Due to its most severe operation conditions, the petrochemical FCC demands even more concerns related to corrosion phenomena, especially related to mechanisms associated with higher temperatures in the reaction sections.

15.5 CORROSION MANAGEMENT IN HYDROPROCESSING UNITS

Hydrotreatment technologies aim to remove contaminants from oil fractions, especially sulfur and nitrogen, to reduce SOx and NOx emissions by derivatives, as shown here:

$$R\text{-}CH=CH_2 + H_2 \rightarrow R\text{-}CH_2\text{-}CH_3 \text{ (Olefin Saturation)}$$
$$R\text{-}SH + H_2 \rightarrow R\text{-}H + H_2S \text{ (Hydrodesulfurization)}$$
$$R\text{-}NH_2 + H_2 \rightarrow R\text{-}H + NH_3 \text{ (Hydrodenitrogenation)}$$
$$R\text{-}OH + H_2 \rightarrow R\text{-}H + H_2O \text{ (Hydrodeoxigenation)}$$

Where R represents a hydrocarbon

Hydrotreating is applied in the finishing of the final products like gasoline, diesel, or kerosene or the intermediate step in the refining scheme in refineries to prepare feed charges to other processes like RFCC or hydrocracking (HCC), where the main objective is to protect the catalyst applied in these processes.

The corrosion phenomena in hydroprocessing units can be divided into phenomena associated with high temperatures and associated with low temperatures. The corrosive processes associated with high temperatures are as follows:

- Sulfide corrosion
- Naphthenic acid corrosion
- High-temperature hydrogen attack (HTHA)
- Hydrogen embrittlement

The sulfide corrosion process is quite common in oil refineries and occurs due to the degradation of steel through the reaction of iron with sulfur compounds contained in the unit's feed streams, usually above 260°C. In the case of regions of the unit without the presence of hydrogen, the application of steels containing 5–12% chromium is considered robust to ensure an adequate, useful life for process equipment. The higher the chromium content in steel, the greater the resistance to corrosion by sulfide. The addition of silicon to steels applied to hydrotreating units also contributes to reducing the rate of corrosion by sulfide.

In the regions where hydrogen is present, the corrosive process is even more severe and follows a different mechanism from the one in the absence of this compound due to the reducing nature of the atmosphere (presence of H_2S). In these cases, austenitic stainless steels are used, especially in severe process conditions, such as deep hydrotreatment or hydrocracking units.

Naphthenic corrosion generally occurs in units subjected to the processing of currents with high acidity. The control parameter used is the TAN, which is defined as the amount of KOH required to neutralize 1 gram of the intermediate. Usually, a number of total acidities above 0.30 mg KOH/g and higher temperatures at 240°C can indicate the possibility of corrosion by naphthenic acids. In addition to these factors, the turbulent flow regimes contribute to accelerating the corrosion rates in this case. The control of the total acidity of crude oil can indicate the need for control measures regarding the naphthenic corrosion in hydroprocessing units. However, the combination of high acidity with reduced levels of sulfur in crude oil and consequently of the intermediate currents can lead to greater severity of the corrosive process in this case since the low sulfur content limits the formation of the protective layer of iron sulfide (FeS). Combating the mechanism of naphthenic corrosion usually involves the addition of molybdenum to steel, providing greater resistance to attack by naphthenic acids.

The attack by hydrogen at high temperature (HTNA) causes the reduction of the mechanical resistance of the steel due to the formation of flaws in the material structure caused by the reaction between hydrogen and carbon, according to the following mechanism:

$$8H + C + Fe_3C \leftrightarrow 2CH_4 + 3Fe \qquad (15.13)$$

The presence of methane in the structure induces the formation of cracks in the steel structure and can lead to premature failures. In this case, preventive actions require the selection of adequate alloy steel (Cr, Mo, and V) and operation within the recommended pressure and temperature parameters, operating below the Nelson curve, which is a relation between operating temperature and hydrogen partial pressure.

The hydrogen embrittlement mechanism involves the contamination of steel by molecular hydrogen (H_2), leading to the risk of failure, especially in periods of instability such as stopping and starting the processing unit. The procedure for stopping and starting hydrotreating units is normally carried out at the lowest possible pressure, as well as controlling the cooling or heating rate of the reaction system.

Corrosion Management in Refining Assets

Due to the need for high temperatures, the above mechanisms are usually of concern in the load and reaction heating sections.

Corrosion processes associated with low temperature normally occur in the hydrogen separation and recycling sections of the hydrotreating units. The main mechanisms are as follows:

- Salt deposition
  1. NH_4Cl, HCl
  2. NH_4F, HF
  3. NH_4HS
  4. H_2S fragilization
- Stress corrosion during shutdowns
  1. Attack by polythionic acids
  2. Chloride attack

The corrosion process resulting from the salt deposition is associated with the contaminants present in the charge currents of hydroprocessing units. The processing of currents containing sulfur and nitrogen leads to the formation of H_2S and NH_3 in the unit's outlet currents as recycled gases. Such gases can be combined to produce NH_4HS, according to Reaction 15.14.

$$NH_{3(g)} + H_2S_{(g)} \leftrightarrow NH_4HS_{(s)} \tag{15.14}$$

The concentration of salts formed depends on the content of contaminants in the unit's load, so units dedicated to the hydrotreating of bottom barrel streams tend to show more severe processes of corrosion by this mechanism.

Refiners typically use water wash injection to minimize the deposition of corrosive salts at key points on the unit, such as the inlet pipes of the unit's separation section and replacement hydrogen. Figure 15.4 indicates the water wash injection point normally used in hydrotreating units.

The quality of the water wash is a relevant factor in the control of corrosion and salt deposition. Some refiners use boiler water or steam condensate for this service. However, due to the high cost of demineralized water, the application of rectified water from acid water rectification units is more common. In this case, the use of rectified waters from deep conversion units, such as delayed coking and FCC, should be avoided, which may contain cyanides that accelerate the corrosive process. Another relevant factor is the temperature of the water wash. The water needs to be injected at a temperature below the salt deposition temperatures. The salt deposition temperatures vary with the severity of the hydrotreating units once the partial pressure of the salts is higher than high hydroprocessing pressures.

The embrittlement by wet H_2S can lead to the embrittlement of carbon steel due to the presence of H_2S or NH_4HS in the presence of water. In this case, the correct selection of materials and post-welding stress relief treatment can mitigate the risk of failure.

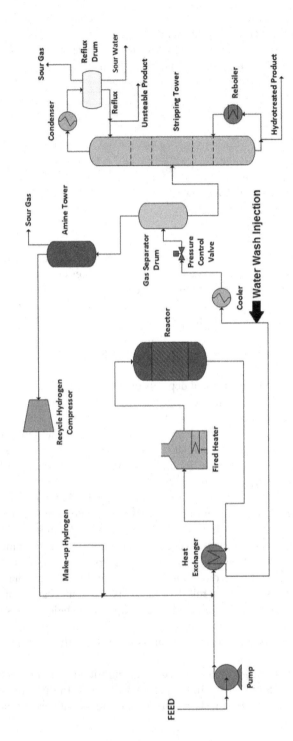

FIGURE 15.4 Water Wash Injection Point in Hydrotreating Units

The embrittlement by polythionic acids is common in cargo-heating and effluent-cooling systems in the reaction systems. In this case, contamination with oxygen and water during maintenance stops leads to the formation of acids and the risk of corrosion under stress. The appropriate selection of stainless steel (stainless steel 321 and 347) and the inertization of systems with hydrogen or nitrogen during maintenance stops can minimize the risks of corrosion by polythionic acids.

Chloride corrosion is a particular concern for hydrotreating units that consume hydrogen from catalytic reforming units. These units consume chlorine in the process of regenerating the acid function of the catalyst, and there may be contamination by chloride (HCl) in the hydrogen produced, leading to contamination of the hydrotreatment unit. Austenitic stainless steels are very susceptible to failure due to chloride attack, and their application should be avoided in regions susceptible to chloride contamination and under temperatures above 60°C.

Like a flexible refining technology, FCC units have a highlighted role in the future of the downstream industry, especially considering the transitive period, where FCC units can help to supply the demand for petrochemicals without a shortage of transportation fuels. In this sense, adequate monitoring and control of the corrosion process in these units are fundamental to ensuring the competitiveness of the players in the downstream sector.

The availability of fundamental processing units like FCC and hydrotreating is a key issue for refiners, as mentioned previously, and the control of the unit's corrosive process plays a key role in ensuring reliable and safe operations of the refining hardware, especially hydroprocessing units dedicated to the treatment of bottom barrel streams that operate under more severe process conditions and under a higher content of contaminants. In the current scenario, the relevance of these units to the profitability of refiners has become even more critical in view of the greater severity in the treatment of bottom barrel or intermediate streams to meet the new quality requirements for marine fuel oil (bunker) in compliance with IMO 2020.

BIBLIOGRAPHY

1. Ramanathan, L.V. *Corrosion and Its Control*. 1st edition, Hemus Press, 1978.
2. Otzisk, B., Magri, F., Achten, J., Halsbergue, S. *Preventing Ammonium Salt Fouling and Corrosion*, PTQ Magazine, 2017.
3. Zhang, W. Evaluation of Susceptibility to Hydrogen Embrittlement—A Rising Step Load Testing Method. *Materials Sciences and Applications*, 2016 (7) 389–395.
4. Revie, R.W., Uhlig, H.H. *Corrosion and Corrosion Control—An Introduction to Corrosion Science and Engineering*. 4th edition, Jhon Wiley & Sons Press, 2008.
5. API 571/2011. *Damage Mechanisms Affecting Fixed Equipment in the Refining Industry*, American Petroleum Institute, 2011.
6. Harston, J.D., Ropital, F. *Corrosion in Refineries*. 1st edition, CRC Press, 2007.
7. Speight, J.G. *Oil and Gas Corrosion Prevention—From Surface Facilities to Refineries*. 1st edition, Elsevier Press, 2014.

16 Energy Management and the Sustainability of the Downstream Industry

16.1 INTRODUCTION AND CONTEXT

The increasing necessity to reduce the environmental impact produced by fossil fuels have created a trend of decarbonization of the energetic matrix at a global level, creating a new challenge to the crude oil production and processing chain. The downstream industry is responsible for producing energy, but a significant part of this energy is applied in the refining processes, and unfortunately, this energy is supplied through carbon emissions. Considering this fact, minimizing the energy intensity of the crude oil refining processes is fundamental to ensuring the sustainability of the downstream business.

In this sense, there are two driving forces responsible for leading the players of the downstream industry to deeply consider the energetic management in their strategic business plan and the environmental and economic factors. Nowadays, the strategy of the companies needs to look for compliance with the sustainability triad: people, planet, and profit, in this order.

As presented in Chapter 7 of this book, the profitability of a crude oil refinery is directly proportional to its capacity to add value to processed crude oil, aiming to maximize the production of high-added-value streams and derivatives. Equation 16.1 presents a simplified concept of the liquid refining margin.

$$Liquid\ Refining\ Margin = \sum_{i}^{n}(Di \times vi) - Pc - (Fc + Vc) \quad (16.1)$$

The first term in Equation 16.1 corresponds to the obtained revenue through the commercialization of crude oil derivatives, represented by the sum of the product between the derivative market value and the volume or weight commercialized. As aforementioned, the profitability or refining margin is directly proportional to the refinery capacity to add value to the processed crude slate. The maximization of higher-added-value derivatives leads to the maximization of the first term in Equation 16.1.

In Equation 16.1, the term Pc corresponds to the acquisition cost of crude oil. The refiners normally don't have control over the market, and the acquisition cost of petroleum is controlled by the geopolitical scenario and the international market. The cost reduction in crude oil acquisition can be achieved through the processing of heavier crude oil, which has a relatively low cost. However, due to the lower distillate

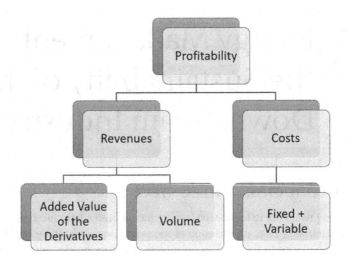

FIGURE 16.1 Simplified Composition of Profitability for Crude Oil Refineries

yield and higher contaminant content, the processing of heavier crude slates requires refining configurations with higher conversion capacity, raising the operational costs.

The fixed and variable costs ($Fc + Vc$) represent the operational costs of refiners. Figure 16.1 shows a simplified scheme for the profitability in the downstream industry, highlighting the refiners' cost composition.

Considering that the energy consumption represents over 60% of the total costs of a typical crude oil refinery, the capacity to optimize and reduce the wastage of energy is a key factor in the competitive game in the downstream industry.

16.2 SIMPLE AND AVAILABLE ALTERNATIVES TO ENERGY OPTIMIZATION

In a typical crude oil refinery, it's possible to identify two basic ways to improve profitability through energy optimization:

1. Consumption reduction: Normally, this way involves a direct impact on the production yield and quality of the derivatives or in the auxiliary processes, so the simple reduction in the energy consumption can be a difficult alternative.
2. Cost reduction of the energy: This way involves actions capable of optimizing the energy consumption and minimizing the energy wastages through reliability improvement. These actions will normally represent a reduction in the global energy consumption of the refinery.

Among the optimization opportunities, some questions must be answered, like when is it advantageous to use thermal or electric energy, what is the best operation scenario to change electrical equipment to turbines, and what is the best time to maximize the internal power generation in the refinery?

These questions can be answered using adequate simulators considering key information like natural gas prices, equipment availability, and reliability, as well as the operating threats of key processing units like FCC.

Among the consolidated techniques to optimize energy consumption, we can note the use of Pinch analysis to reduce the necessity of hot utilities in the processing units and maximize the energy recovery from the process. This alternative is especially attractive in crude oil distillation, FCC, and delayed coking units due to the amount of energy applied to the processes.

The use of equipment load management systems, speed monitoring for rotating equipment, monitoring thermal efficiency, controlling and monitoring the flare gas emissions, and measuring the flow rate of gases discharged to the flare system can help identify bad actors and promote effective actions to optimize the power consumption in the process plant.

Some alternatives to reduce the energy need in crude oil refineries involve some capital investments in revamps or new designs. A wide energy optimization action has been the revamp of fired heaters and furnaces to improve efficiency and minimize greenhouse gas emissions.

The crude oil distillation unit has the largest energy consumption in a crude oil refinery, and there are many optimization opportunities in this processing unit. Establishing an adequate overflash can save a great amount of energy in the fired heaters. For processing units with higher processing capacity, the use of preflash configuration can help the refiners reduce the energy intensity of the downstream steps of the process. Figure 16.2 presents a crude oil distillation unit relying on a preflash column.

An evolution of the preflash configuration is the progressive distillation technology developed by Technip. This concept considers that the lighter fractions are superheated in the conventional process, leading to energy wastage. To avoid this effect, the progressive distillation concept applies a series of distillation towers focusing on supplying only the required energy to promote the desired separation. Figure 16.3 presents a basic process flow with the concept of progressive distillation.

The separation processes are intensive in the energy consumption, and the optimization of the internals of towers and separating vessels can help the refiners minimize energy consumption, and these actions can be an attractive alternative to revamp the processing units. The use of low pressure drop internally and dividing walls in distillation columns are effective actions in this sense.

Optimization and design actions can offer good results, but it's necessary to ensure the reliability and the availability of the refining hardware before performing these actions. Two of the main issues of the energy system in a crude oil refinery are steam leaks and the recovery of steam condensate. Maximizing steam condensate recovery and minimizing steam leaks seem obvious but are fundamental and sometimes

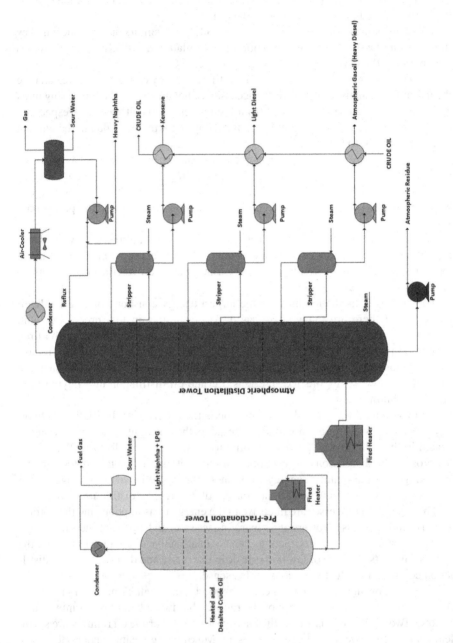

FIGURE 16.2 Crude Oil Distillation with a Preflash Column

Energy Management and the Sustainability of the Downstream Industry 261

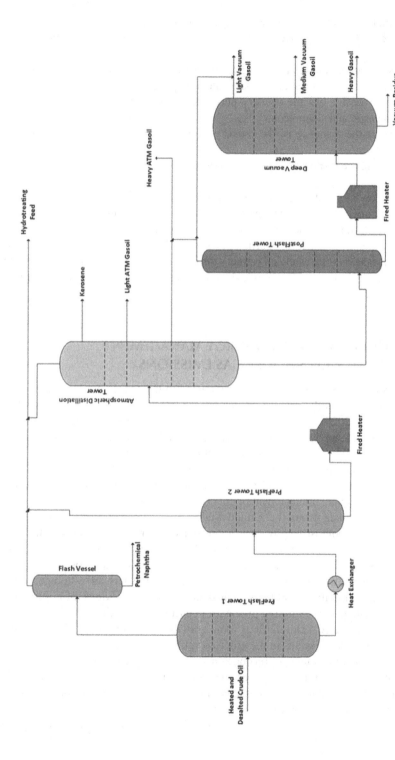

FIGURE 16.3 The Concept of Progressive Distillation

underestimated by some refiners. Some questions need to be answered as a diagnostic step to reveal the real concern of these questions to a refinery management staff.

1. Has a specific maintenance program been established for steam traps? What is the frequency of review of that evaluation program?
2. How proactive has the refinery personnel been in managing and solving condensate and steam leaks?
3. What is the response of the refinery to managing compressed air system leaks?
4. What are the responses if insulation systems are subpar?

Another key moment to reaching an optimized energy system in a crude oil refinery is the planned shutdown. The participation of process engineers in the planning steps of a turnaround of the processing units is fundamental to helping the maintenance team identify and solve inefficiency points. Also, it's fundamental to understand the scenario of the crude oil refinery during the turnaround. Which processing units will stop? How can we optimize the operation of the processing units that will keep in operation?

16.3 THE IMPACT OF THE ENERGY MANAGEMENT ON THE GREENHOUSE GAS EMISSIONS

More relevant than the profitability issues related to the energy management of the crude oil refineries is the relation of greenhouse gas emissions. According to data from RITCHIE (2020), the energy sector is responsible for the major part (more than 70%) of the world's greenhouse gas emissions, the industrial sector is responsible for more than 24% of this total, and the crude oil industry is responsible for a significant amount of these emissions. In this sense, achieving more sustainable operations is essential to control and minimize the carbon emissions in crude oil refineries, and this relies on adequate management of energy consumption.

According to data from the United States Environmental Protection Agency (EPA, 2010), the combustion processes are responsible for more than 60% of the CO_2 gas emissions in a typical crude oil refinery, followed by the burn of catalyst coke in FCC units with 23% and the hydrogen production plants with close to 6%. The remaining contribution is shared with other processes, like derivative storage, sulfur recovery plants, and flaring systems. This data reveals the relevance of minimizing the energy wastage in the refining processes to ensure the sustainability of the refining business. Figure 16.4 summarizes some available actions capable of minimizing the greenhouse gas emissions in the crude oil refineries.

The energy management actions became essential to the sustainability of crude oil refining players and needed to be an essential part of any strategy in the downstream industry both for environmental impact and economic competitiveness between the players. As presented previously, the sustainability of the downstream business is strongly dependent on the adequate energy management of the refining assets.

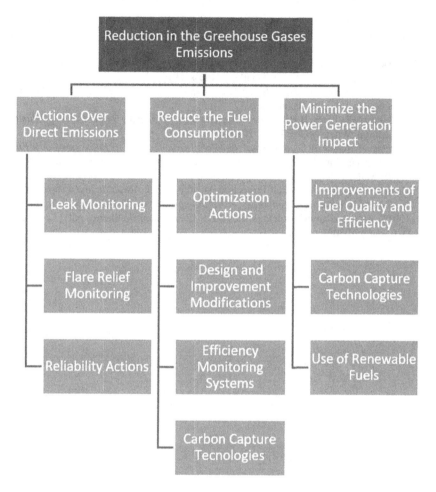

FIGURE 16.4 Available Actions to Minimize Greenhouse Gas Emissions in Crude Oil Refineries

Sustainability is not a choice but a survival question in the downstream industry. Adequate energy management actions are fundamental to ensure not only competitive economic positioning but, even more importantly, the reduction of greenhouse gas emissions, which is essential to allow the achievement of a low-carbon economy. The recent trends of the downstream industry, like renewables coprocessing and higher demand by petrochemical intermediates, reinforce the necessity of energy optimization actions to minimize the energy intensity of downstream operations.

BIBLIOGRAPHY

1. Karatas, Z., Turkoglu, S. *A Refinery's Journey to Energy Efficiency*, PTQ Magazine 2016.
2. Rikhtegar, F., Sadighi, S. *Optimization of Energy Consumption*, PTQ Magazine 2015.

3. Concawe. *EU Refinery Energy Systems and Efficiency: Conservation of Clean Air and Water in Europe 2012*, www.concawe.eu/wp-content/uploads/2017/01/rpt_12-03-2012-01520-01-e.pdf (Accessed on 01 March 2022).
4. Fawkes, S. *Energy Efficiency: The Definitive Guide to the Cheapest, Cleanest, Fastest Source of Energy*. 1st edition, Routledge Press, 2013.
5. Kaiser, V. *Industrial Energy Management: Refining, Petrochemicals and Gas Processing Techniques*, Editions Technip, 1993.
6. Rossiter, A.P., Jones, B.P. *Energy Management and Efficiency for the Process Industries*. 1st edition, Wiley-AIChE Press, 2015.
7. Ritchie, H. *Sector by Sector: Where do Global Greenhouse Gas Emissions Come From?* Our World in Data, 2020, https://ourworldindata.org/ghg-emissions-by-sector (Accessed on 01 March of 2022).
8. United States Environmental Protection Agency (EPA). *Available and Emerging Technologies for Reducing Greenhouse Gas Emissions from the Petroleum Refining Industry*, 2010, www.epa.gov/sites/default/files/2015-12/documents/refineries.pdf (Accessed on 01 March of 2022).

Index

A

alkylation, 8, 9, 56
API grade, 1
API separator, 169
asphalt, 12
asphaltene, 12, 27, 30, 85
ammonia, 224–226
aromatic residue, 44
aromatics, 52–54
aromizing process, 51
atmospheric distillation, 15
atmospheric residue, 88

B

bender treating, 152
benzene, 28, 49–52
biofuels, 195–200
biological treating, 170
bioreactor, 170–171
BMCI, 45
boiling point, 19, 44, 99
bottom barrel, 84, 100
bunker, 3, 12, 90, 133
business strategy, 229

C

carbon residue, 32, 102
catalyst, 40–42, 50–52
catalyst deactivation, 32, 221
Catalyst Group, the, 186, 187
catalytic conversion, 34
catalytic cracking, 35
CCR platforming, 51
Chevron Lummus, 88, 98, 131, 180, 182
Claus process, 156
clean hydrogen, 141
coke, 26–30
coke deposition, 19, 40, 81, 129, 220
coker naphtha, 77
corrosion, 241
crude oil, 1–7
crude oil derivatives, 7
crude oil distillation, 15–19
crude oil to chemicals, 178

D

deasphalting, 30
decarbonization, 194, 257
delayed coking, 35
desulfurization, 88
dewaxing, 83
dienes control, 81
diesel, 82, 199
diesel hydrotreating, 82
discounted crudes, 177, 230
distillation, 15, 17

E

ebullated-bed hydrocracking, 98, 182–184
economic sustainability, 103, 109
emissions standards, 262
energy management, 257, 262
environmental impact, 26, 41, 155
ETBE, 66
etherification, 66–69
ethylene, 181, 197
extra-heavy crude, 2

F

FCC units, 35–39
flexicoking, 30, 123
fluid catalytic cracking, 39
fractionation section, 52
fractionation tower, 18
fuel gas, 7
fuel oil, 3
furfural extraction, 107
furnace cracking, 17, 23

G

gas oil streams, 11, 20
gasoline, 7, 28, 43, 56
gas to liquids, 219
global energetic matrix, 141
gravity, 1, 129
greenhouse gas emissions, 262

H

HDS, 93
heavy crude oil, 2
HF acid, 56
H-Oil process, 148, 182
HVO, 199
hydrocracking, 91
hydrodealkylation, 55
hydrodesulfurization, 73
hydrogen, 139
hydrogen management, 142–144
hydroprocessing catalyst, 99
hydroprocessing technologies, 73
hydroskimming refinery, 121
hydrotreated naphtha, 80
hydrotreated vegetable oil, 199
hydrotreating, 77, 79, 82

I

ignition quality, 10
integrated players, 173, 230, 236
integrated refining schemes, 177
integration of refining and petrochemical assets, 47, 190
isobutene, 58
Isomar process, 55
isomerate naphtha, 61–63
isomerization, 61
isooctane, 9
isoparaffinic hydrocarbons, 56
isopropyl-benzene, 58

J

jet fuel, 9–10
JFTOT, 9

K

kerosene, 4, 10, 39, 86

L

LC-fining process, 131, 182
LCO, 36
LHSV, 83
light paraffin, 69
liquid-liquid extraction, 30–32
low sulfur bunker, 3
LPG, 6, 7
lubricating refineries, 105

M

MEA, 160
metal content, 20, 36, 45, 191

methane, 137
methanol to olefins, 214
mild hydrocracking, 99
molecular management, 54
molecular sieves, 63
MTBE, 63

N

naphtha alkylation, 56
naphtha hydrotreating, 77
naphtha isomerization, 60
naphthenic corrosion, 241
natural gas, 6, 46, 137
needle coke, 25
Nelson index, 125
nickel, 40, 42, 86
nitrogen content, 4, 93, 129

O

octane, 9, 43
oily sewer, 167
olefin metathesis, 214
olefins condensation, 64
operating costs, 2, 4, 142
oxygenated additives, 68

P

paraffin dehydrogenation, 69
plastics recycling, 192
platforming reforming process, 51
platinum, 49, 62, 212, 224
porter competitive forces, 230
propane dehydrogenation, 211
propylene gap, 203
p-xylene, 54–56

Q

quality requirements, 4, 7, 44, 144

R

refining configurations, 119
refining integration, 237
refining margins, 34, 84, 119
Reid vapor pressure, 7, 9, 61, 64, 82
renewable fuels, 195
renewable hydrogen, 141
renewables coprocessing, 194
residue fluid catalytic cracking, 39
residue upgrading, 84, 190
resins, 30
ROSE technology, 32

Index

S

salt content, 4
Saudi Aramco, 178, 179, 187
shift reaction, 137
shot coke, 27
soaker visbraking, 24
solvent deasphalting, 30
sour water stripping, 155
sponge, coke, 27
STAR dehydrogenation process, 71
steam cracking, 211
stormwater sewer, 167
sulfur recovery, 156, 160
sweetening, 151
S-Zorb process, 81

T

tail gas, 160–161
TAME, 66
thermal conversion, 23
toluene, 28, 52, 55
topping refinery, 119

total acid number, 4, 101, 241
transfer line, 81
treating process, 4, 30, 147

U

uniflex process, 88, 131, 182, 184
UOP, 32, 39, 51, 55
used lubricating recycling, 111

V

vacuum distillation, 11, 19, 21, 32
vanadium, 11, 40–42, 85
visbreaking, 23

X

xylenes, 52, 54

Z

zeolite, 35, 42–43, 207

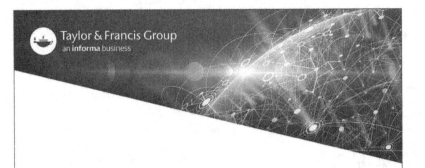

Printed in the United States
by Baker & Taylor Publisher Services